Physical Models for Semiconductor Devices

Physical Models for Semiconductor Devices

John E. Carroll

Fellow of Queens' College,
Lecturer in Department of Engineering,
University of Cambridge.

Edward Arnold

First published 1974
by Edward Arnold (Publishers) Ltd
41 Bedford Square, London WC1B 3DQ

Boards Edition ISBN: 0 7131 3307 4
Paper Edition ISBN: 0 7131 3308 2

Reprinted with corrections 1980

Printed in Great Britain at The Pitman Press, Bath

Contents

Preface

It is the author's experience that many students find it easier to tackle the more abstract problem once they have seen its practical applications and so have the subject more in perspective. Even the student with theoretical ability sometimes uses the mathematics mechanically without fully understanding the physical concepts. In this book the practical features and physical models of semiconductor devices are presented as descriptively as possible bearing in mind that the level of study spans the final years of a degree course and the first year of a postgraduate course. A measure of mathematical competence in solution of differential equations and algebraic manipulation is therefore assumed; familiarity with the elements of both kinetic theory and statistical mechanics is also desirable. In order to make the physical aspects as clear as possible, the only aspects of quantum mechanics used in the book are those involving wave-particle duality.

The high degree of cooperation between the physical, chemical and engineering sciences that is needed for the practical development of semiconductors must be appreciated, and this cooperative technology has therefore been made an important feature in this book. The devices chosen for discussion will not only emphasize the cooperative technology but will have interesting physics and practical importance. Clearly it is not possible to include all such devices. Amongst those that are omitted are devices using coupled-wave phenomena, acoustic waves and helicon waves; the author sees these in a distinctive class requiring special treatment outside the scope of this book. For similar reasons the problem of noise in semiconductor devices has been omitted; rather than present the student with a superficial treatment of noise, more space has been given to the mechanisms at larger signal levels—which must be understood before considering the complexities of noise.

The first chapter is a background wash for the details of the picture to come. It is followed by two basic chapters—one on materials and manufacturing techniques and one on conduction processes. While Chapter 2 is fairly light reading, the chapter on conduction and equilibrium processes will repay additional study after later chapters have been read. The object has been to explain carefully about electrical neutrality, when it is valid and when it is violated, introducing the concepts of dielectric relaxation time and Debye length. Here also are introduced the Fermi level, diffusion and recombination—these being the mechanisms by which equilibrium statistics are achieved.

Chapters 4 and 5 consider the classical transistor physics of p-n junctions and the related devices. Chapter 4 concentrates on the junction diode, emphasizing the different mechanisms which can limit or control the current flow through a p-n junction; the step-recovery diode, varactor and tunnel diode have also been covered here. Chapter 5 presents an important innovation—a model of an 'ideal' transistor. It is hoped that this will make the key physical processes much clearer to the student than the older teaching about diffusion transistors. The central importance of bipolar transistors justifies a thorough study, so an account of the high-frequency operation is also included, using the elegant ideas of the charge-storage models for the transistor. The chapter ends with accounts of the field-effect transistor and semiconductor control rectifier.

While there are many intermediate or advanced books that help the theoretically inclined student to understand band gaps, effective mass and collision theory, the author considered that a more elementary account could be of help in explaining the physical concepts. An appreciation of the wave nature of the electron, the periodic structure of a crystal and some algebraic manipulation elucidates many features of energy bands in crystal. Chapter 6 covers this and paves the way for a better understanding of optical effects in semiconductors (Chapter 9), electron transfer or Gunn effect (Chapter 8) and the thermionic emission theory of Schottky barrier diodes (Chapter 7). This latter theory is more modern than the diffusion theory and is more appropriate for semiconductors with a high mobility. Chapter 7 also covers some basic theory required in modelling MOS devices including the recent charge coupled devices.

Chapter 8 gives a short account of the use of energetic electrons in semiconductors and, in particular, their role in generating microwaves. The avalanche oscillator and the Gunn effect are also discussed. This has been a rapid area of growth and its inclusion is essential to give a proper perspective.

Growth has been equally rapid in the development of optical devices such as light emitting diodes, X-ray and γ-ray detectors, solar cells and television cameras. Topics like these are covered in Chapter 9. To give a sound backing to the theory, this chapter is opened with the Einstein treatment of spontaneous and stimulated emission. Unlike the usual simplified treatments, the full Fermi-Dirac statistics have been retained for the electrons and this helps the discussion of the injection laser at the end of the chapter.

The appendices extend the theory with more specialized topics. The final appendix emphasizes the importance of *magnitudes* in deciding which aspect of the physics dominates the device model. Change these magnitudes and different aspects become important and the model for a device can radically change.

The problems at the end of each chapter are intended to make the student think through additional areas for himself. Working carefully through a few unseen problems will do much more to develop a student's confidence than any number of automatic calculations. The book has been written in SI units to encourage the reader to conform to the international standard, though most practising semiconductor experts still think in cgs units!

It is hoped that the student will finish the book with a feeling for the activity that has gone into the development of so many semiconductor devices. As is usual in any book, one finds that the material grows from discussions with colleagues and friends; it is often difficult to identify the source of a particular view. It is hoped that those who have spent time talking to the author, showing him around their factories and their development or research laboratories, will appreciate how much he has valued their efforts. Students asking the difficult questions that test whether one knows the subject as well as one thought have also been of much help.

Special acknowledgements

The author would like to make special acknowledgements to Professor A. H. W. Beck and Professor J. Lamb for helpful comments, to Dr C. Gooch for help with aspects of Chapter 9, and to Drs S. M. Sze, W. Fawcett, P. Schagen and V. Heine for permission to use some of their published material.

John E. Carroll
Cambridge 1973

Principal symbols

Several symbols have duplicate meanings and the reader should observe both context and dimensions.

A	area, crystal length
B	magnetic field
C_s	surface concentration
C_n, C_p	capture cross sections for electron and hole traps
d	distance, length
D	diffusion coefficient for impurities
D_n, D_p	electron and hole diffusion coefficients
e	electronic charge ($1\cdot60 \times 10^{-19}$ C)
e	exponential
E	electric field (sometimes with a suffix e.g., E_p for peak field)
$\mathscr{E}, \mathscr{E}_c, \mathscr{E}_v$	Energy, conduction band energy, valence band energy
$\mathscr{E}_f$	Fermi energy
f	fraction of states occupied, frequency
$F(\mathscr{E})$	distribution function of energy
g, g_d, g_m	conductance, diode conductance, mutual conductance
G, G_{opt}	generation rate, optical generation rate
h_i, h_o, h_r, h_f	h-parameters (section 5.1)
h_{fe}, h_{fb}	common emitter and common base forward current transfer ratio
h	Planck's constant [$6\cdot625 \times 10^{-34}$ J s]
$\hbar$	$h/2\pi$
i	current
I	current (often with suffix to denote special value e.g., I_s reverse saturation current)
J	current density
k	Boltzmann's constant [$1\cdot38 \times 10^{-23}$ J/K]
k	$2\pi/\lambda$ propagation constant
k	fraction of charge stored that is recoverable
k_1, k_2	fractions of charge stored supplied from emitter and collector current components in transistor (section 5.2)

K	propagation constant $(2\pi/\lambda)$, degrees Kelvin
K_1, K_2	fractions of charge supplied in dynamic state to stored base charge in transistor (section 5.6) length
L	length
L_D	Debye length $(L_D^2 = kT\varepsilon_r\varepsilon_0/e\rho_0)$ (section 3.4)
L_p, L_n	Diffusion length for holes and electrons (section 4.3)
m	integer, mass (electronic mass $m = 9\cdot11 \times 10^{-31}$ kg)
m^*	effective mass
n	integer, refractive index, electron concentration
n_i	intrinsic value of electron or hole concentration
N_v, N_c	effective densities of states for valence and conduction bands
N_A, N_D	acceptor and donor concentrations
p	hole concentration
P	fixed value of hole concentration, photon density
q	charge
Q, Q_s	charge, stored charge
R	reflection coefficient (Chapter 6), recombination rate (Chapter 3)
R_e	equilibrium recombination rate
R, r, R_L	resistance, load resistance
r, T	time
T	transmission coefficient (Chapter 6)
T, T_e	temperature, electron temperature
v, v_s	velocity, scattering limited velocity
V, V_a, V_f, V_{be}	voltage, applied voltage, forward voltage, base-emitter voltage
W	width
x, X	distance
X	reactance
Y	admittance
Z	atomic number
α	ionisation rate (section 8.3)
α	common base current transfer ratio (Chapter 5)
α	absorption coefficient for photons (section 9.4)
β	common emitter current transfer ratio
ε_r	relative permittivity
ε_o	permittivity of free space [$8\cdot854 \times 10^{-12}$ F/m]
η	efficiency, reciprocal length (Chapter 5)
λ	wavelength
μ	mobility
μ_0	permeability of free space [$4\pi \times 10^{-7}$ H/m]

π	pi [3·1416]
ρ	charge density, resistivity
σ	conductivity
$\tau, \tau_r, \tau_{rn}, \tau_{rp}$	time interval, recombination time, recombination time for electrons and for holes
τ_e, τ_m, τ	energy relaxation time, momentum relaxation time, free time between collisions
ϕ	voltage, work function, normalised voltage, phase angle
χ	electron affinity
ψ, ψ_s	electron quantum wave function, electric potential, surface potential
ω	angular frequency ($2\pi f$)

Some orders of magnitude:

Material	effective masses in terms of free electron mass		mobility m^2/V s	permittivity $\varepsilon/\varepsilon_0$	band gap eV
GaAs	electrons	0·07	0·8	12·5	1·4
	holes	0·5	0·05		
Si	electrons $m_\parallel$	1·0	0·15	12	1·1
	$m_\perp$	0·2			
	holes	0·5	0·05		
Ge	electrons $m_\parallel$	1·6	0·35	16	0·7
	$m_\perp$	0·08			
	holes	0·3	0·19		
InP	electrons	0·07	0·6	14	1·3
	holes	0·4			

1

Semiconductor Devices

Almost every reader will have a transistor radio set; a few will have pocket electronic calculators displaying the answers in bright arrays of light-emitting semiconductor diodes. Some will be caught speeding in their cars by 'transistorized' radar sets that are too small to notice by the side of the road. For the romantic evening one can buy a semiconductor light-dimmer to replace the switch on the wall. The toy train reverses its direction through the action of a semiconductor diode rectifying the a.c. mains supply. Computers containing myriads of semiconductor elements store more and more information but get smaller, and even cost less; they also talk to each other by impulses of light sent out from semiconductor devices and received by other semiconductor devices. Many readers will have built amplifiers for their record players by merely connecting the power to an integrated semiconductor circuit. These random thoughts are jotted down to illustrate the broad impact of semiconductor devices. To the author, systems and instruments which use these devices are made all the more interesting and understandable by an appreciation of a physical model for the underlying device. The aim of this book is to provide a foundation of physics and device models for a number of the important high-quality semiconductor devices that are available.

The reader should already have met the concepts of n-type and p-type semiconductor material, conduction of electricity being by the motion of negative charges (electrons) in the former and positive charges (holes) in the latter. A number of devices can be readily understood merely by remembering that positive charges are attracted towards points of negative potential and negative charges towards points of positive potential. Indeed, the simplest form of rectifying action in the p-n junction can be understood in these terms (Fig. 1.1). For many practical devices made from combinations of p-n junctions this simple physics can often help the student get the polarities correct. Thus 'p-section positive' is a well known mnemonic for the easy direction of current flow in a semiconductor junction diode. More complex models are soon required to explain subtle effects such as charge storage in

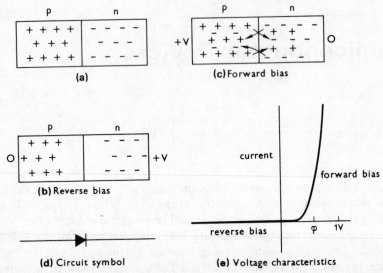

Fig. 1.1 The p-n junction. Material with positive mobile charge carriers and negative mobile charge carriers (a) shown with reverse bias (b) pulling charges apart; in this case there is no source of supply for the charges and current cannot flow. On forward bias (c) the p-electrode can supply the positive charge carriers and the n-electrode the negative charge carriers; current can flow, with the electrons moving towards the positive voltage terminal and the holes in the opposite direction. The arrow on the circuit symbol (d) always shows the direction of motion of positive charge carriers when the p-section material is positive in voltage in respect to the n-material (forward bias). The characteristics are shown in (e). Note that good conduction does not occur until typically around 0·5 V for most semiconductor diodes.

the diode. Quantum effects can change the characteristics of diodes and the 'tunnel' diode (Fig. 1.2) was one of the first semiconductor devices to give a useful 'negative resistance'. This refers to the portion of the characteristic where $dV/dI = -R$, a negative value. One finds that a negative resistance can give out power rather than absorbing it like a positive resistance and the concept is an important one in many semiconductor devices used for generating high frequency oscillations.

The basic physics of p-n junctions is pre-requisite knowledge before discussing groupings of p-n junctions (Fig. 1.3) which form practical control and amplifying devices—the junction bipolar transistor, the junction field effect transistor and thyristors. This basic physics is covered in Chapter 4 leaving Chapter 5 for discussion of the more complex devices.

Not only can the junctions be grouped together to form useful devices but of course the devices can be grouped to form circuits. A visit to a semiconductor device factory, would show that many thousands of diodes and transistors are made simultaneously on discs of material many centimetres in diameter. The material is usually silicon or gallium arsenide. Special

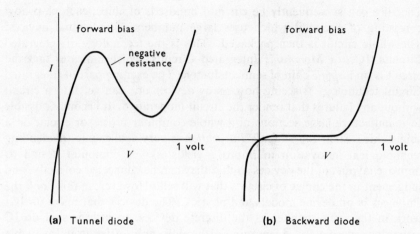

(a) Tunnel diode   **(b)** Backward diode

Fig. 12 Special diodes. The tunnel diode exhibits negative resistance. The backward diode is useful for rectifying very small voltages, much less than 1 V, by use of its high-current characteristic on reverse bias.

applications such as light emitting diodes may require other special materials such as gallium phosphide. Germanium, a popular material at one time, is infrequently used now for reasons that will become apparent later. Nearly every device on such a disc or slice is identical, failures being confined mainly to damaged regions near the edges of the slice. The slice can at that stage be cut into chips containing discrete devices which are then packaged. The number of devices that finally work expressed as a percentage of the number made is the *yield*. Provided the yield approaches 100 per cent, then devices which are grouped together on the slice can be joined by evaporated metal contacts while still on the slice, so forming circuits free from defective devices.

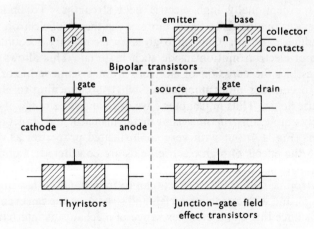

Fig. 1.3   Combinations of p-n junctions.

The slice can subsequently be cut into hundreds of chips, each chip now consisting of a circuit which uses large numbers of individual devices. The whole circuit is then packaged. This is the technology of Integrated Circuits (ICs) or *Monolithic* Integrated Circuits, which indicates that the circuit is on a *single* chip of semiconductor. The exciting part of Integrated Circuit technology is seeing how many devices one can get into a circuit without any failures that render the circuit inoperative. It becomes possible to manufacture large sections of a whole computer processor circuit on a single chip. Such *Large Scale Integration* is very challenging, demanding technological innovation to improve yields, skilled computer design to improve layouts of the devices so that they can be connected correctly, and judgement in the choice of circuits that will sell. However, in this book the emphasis is on device models and physics. The devices that make up ICs work in the same manner as the discrete devices, so little is said on IC technology specifically. A few points of the philosophy of design and available technology are discussed in a section at the end of Chapter 5.

Up to Chapter 6 this book is concerned with traditional transistor physics, though with an emphasis demanded by modern technology. Now one looks at devices on the market; one can find metal–semiconductor junctions, Schottky barriers, that are capable of rectifying (in the conventional low-frequency sense) a.c. signals at frequencies over 10^{10} Hz. Such devices are ousting conventional p-n junctions for many purposes. There are other devices, Metal–Oxide–Semiconductor Transistors (MOSTs), that are used in integrated circuit form to store massive numbers of 'bits' of information for computers, yet occupy mere cubic millimetres of space. A telephone conversation between London and Glasgow may be carried by high frequencies around 10^{10} Hz which are generated from minuscule chips of semiconductor operating at exceptionally high electric field strengths. To understand these new devices one needs to know more of the theory of electron motion inside crystalline material. Chapter 6 gives an elementary account of the wave-nature of electron motion inside such material. This allows one to discuss topics like the emission of electrons from a semiconductor into a metal, and to consider more intelligently what is happening to electrons in high electric fields. Thus in Chapter 7 we can consider a range of devices with relatively simple constructions using metal, insulator and semiconductor combinations (Fig. 1.4) but with very sophisticated properties as devices. In Chapter 8 the effects of high electric fields are considered. Contrary to one's early experience of devices, applying 'too many' volts does not always lead to catastrophic failure. New effects can be found; Ohms law no longer holds (Fig. 1.5), and with certain materials and structures one can even obtain negative resistance through a complex range of mechanisms internal to the device.

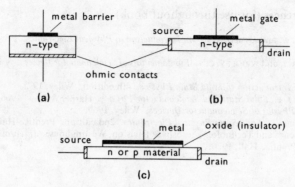

Fig. 1.4 Metal–semiconductor–insulator combinations. (*a*) Schottky barrier diode. (*b*) Schottky gate transistor. (*c*) Metal–oxide–semiconductor field-effect transistor.

We have already indicated that electronic instruments can flash out numbers and messages through tiny light-emitting diodes. Because there are so many telephone and electricity cables under Wall Street, computers just across the road from each other can communicate only by using optical signals generated and *detected* by semiconductor devices. In America again, people can look at their friends on the other end of the telephone through the help of a semiconductor television camera. Satellites high above the earth are powered by silicon solar cells converting the sun's energy into electric current. Because optical devices so clearly have an importance of their own, the book ends in Chapter 9 with a brief account of some of the more significant optical effects in semiconductors and of devices that can be made by using these effects.

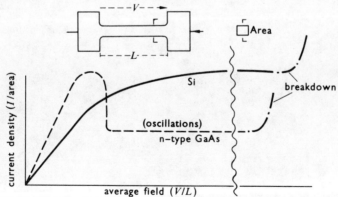

Fig. 1.5 Failure of Ohm's law at high electric fields. The insert shows a typical bar of material, with L a few hundred micrometres, used in early experiments. The contacts were well removed from the central region to try to avoid effects caused by contacts. At high electric fields the current tends to limit; at still higher field strengths the current increases because of 'breakdown' effects.

General references (for use throughout book)

1.1   BECK, A. H. W. and AHMED, A. H. *An Introduction to Physical Electronics*. Edward Arnold, 1968.

1.2   LINDMAYER, J. and WRIGLEY, C. Y. *Fundamentals of Semiconductor Devices*. Van Nostrand, 1965.

1.3   KITTEL, C. *Introduction to Solid State Physics*. 4th edition. Wiley, 1971.

1.4   MCKELVEY, J. P. *Solid State and Semiconductor Physics*. Harper & Row, 1966.

1.5   SZE, S. M. *Physics of Semiconductor Devices*. Wiley, 1969.

1.6   VAN DER ZIEL, A. *Solid State Physical Electronics*. 2nd edition. Prentice-Hall, 1968.

1.7   Staff of Science and Technology Aerospace Division, Westinghouse. *Integrated Electronic Systems*. Prentice Hall, 1970.

2

Semiconductor Materials

2.1 Solids, atoms and electrons

This chapter briefly reviews topics which the reader should have as background knowledge for the study of electronic materials. The detail should be filled in by collateral reading of references 2.1, 2.2 and 2.3, or previous courses. It will be found that most high-performance semiconductor devices are made from single *crystals* of material where the atoms or groups of atoms repeat in a pattern like a three-dimensional wall paper. The smallest unit that can be used to create the crystal pattern by repetition is the *unit cell* (Fig. 2.1). In semiconductors this unit cell is often of a simple form, containing only one or two atoms and having dimensions similar to a 0·5 nm cube. More recent semiconductor devices are beginning to use *amorphous* materials where there is no long range order or repetition, although locally around each atom the pattern of neighbouring atoms is very similar to the crystalline state. However, at present amorphous semiconductor devices are limited to switching-type operations, and indeed the physics is not nearly so well understood as for the crystalline materials. Consequently we shall not discuss 'amorphous' devices in this book but shall concentrate on elements which work with single crystals. The reasons for requiring the close perfection of the periodic pattern of a single crystal may not be apparent to the reader until he has read later chapters; but, we shall nevertheless discuss here the elementary properties and consequences of ideal crystals.

The forces that keep atoms together in a solid arise from the electric charge of the electrons moving around the nucleus of the atoms. Classical mechanics singularly fails to give any insight into this problem, and quantum mechanics is needed. To keep the book short, we have not attempted to combine text on quantum mechanics with discussion of materials. It is assumed that the reader has covered, or is covering in parallel reading, the basic quantum theory (see references 2.3 to 2.7). Thus the reader will be familiar with the wave-particle duality associating a particle of momentum p with a wave of wavelength $\lambda = h/p$, and similarly a wave of frequency f being associated with a packet of energy hf. For electrons one usually finds that the motion is adequately described by the particle approach provided the dimensions

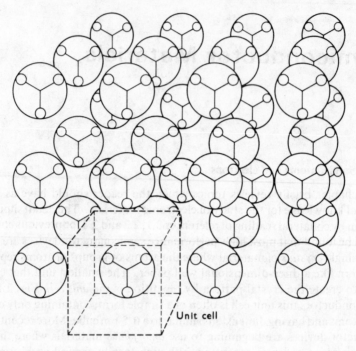

Fig. 2.1 A crystal (schematic).

involved are large compared to a wavelength. However, typical wavelengths for electron waves are comparable to atomic dimensions ($\sim \frac{1}{10}$ nanometre), so that the wave character of the electrons around atoms is most important. Niels Bohr in 1913 presented a theory for the hydrogen atom which utilized the wave nature of the electron. Although the reasoning is no longer acceptable, the answers to some numerical problems are correct (Problem 2.1) and in particular the approach is helpful before starting on a theory of greater complexity. The Bohr theory introduces the concept of a *quantum state* because the permitted electron orbits are limited to those which have a circumference that is an integral number, n, of electron wavelengths. The allowed energies are then $13 \cdot 6/n^2$ eV (the electronvolt, eV, is the energy acquired by an electron falling through 1 volt: $1 \cdot 6 \times 10^{-19}$ J) and the integer n is referred to as the *principal quantum number*. The Bohr theory was discarded when in 1926 Schrodinger introduced his famous wave equation. The solutions to the Schrodinger equation for the hydrogen-like atom may be found in many introductory quantum texts (references 2.3 to 2.8), where it will be seen that other quantum numbers are required to specify a quantum state. These need not concern us in any detail here. The important

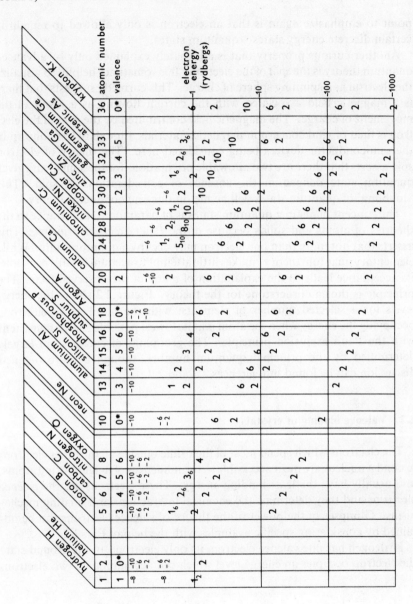

Fig. 2.2 Schematic diagram of energy levels and occupation in ground state for selected elements.

point to emphasize again is that an electron is only allowed to remain in certain discrete energy states—quantum states.

Another curious property that is adequately explained only by advanced quantum theory is the *spin* of the electron. It is sometimes helpful to visualize the electron as a spinning sphere of charge. This correctly suggests that there is a magnetic field associated with the current flowing as a result of the movement of charge. The magnetic field contributed by the spin of the electron is then of vital concern in magnetic materials, though of less concern in the semiconductor devices being considered here. The theory of electron spin asserts that there are two allowable orientations for the spin in any given quantum state defined in the absence of considerations of spin. This 'doubling' of quantum states will be met in our semiconductor theory.

The concept of distinct quantum states is of paramount importance to the theories of atoms and solids because of the *Pauli exclusion principle*. This asserts that not more than one electron may occupy a quantum state. While elementary quantum theory† makes little attempt to account for this principle, its acceptance leads to the explanation of many of the observed facts. The principle is thus a cornerstone for the theory. Figure 2.2 shows the energy levels for a selected number of elements, with the numbers of electrons occupying the energy when the atom is at its lowest possible energy consistent with the Pauli exclusion principle. The groupings of energies are largely determined by the principal quantum number, though a more detailed discussion can be found in references 2.4 and 2.7.

2.2 Valence binding of crystals

The electrons which spend most of their time outside all the other electrons around a nucleus are most accessible to the influences of neighbouring atoms, and usually have the least binding energy. These electrons are the *valence* electrons and they determine the chemical interactions and binding mechanisms. Glimpses of the power of the theory of valence electrons can be obtained by considering specific examples with the help of Fig. 2.2

Hydrogen has one valence electron, its only electron. In the ground state the electron occupies an energy level which can accommodate two electrons

---

† L. I. Schiff,[2.7] for example, discusses the Pauli principle quite well but even so reference has to be made to relativistic quantum mechanics to show that the spin of any particle plays a key role in deciding whether it obeys the Pauli exclusion principle or not. The quantum of spin is measured in h units of angular momentum and the electron has spin $\pm \frac{1}{2}$ in these units. Particles with an odd number of $\frac{1}{2}$ spin, or *half integral spin*, all obey the Pauli exclusion principle, while those of zero or integral spin do not obey the principle. Thus electrons obey the principle but photons of light with no spin do not.

(of opposite spin) without violation of the Pauli principle. This gives hydrogen an ability to lose or accept an electron fairly readily and so makes hydrogen chemically active. In general it is found that when atoms exchange or share valence electrons they have a lower electrostatic energy for their combined electron patterns than when they are separate. This is particularly marked when the complement of quantum states is completed by exchange or sharing. The decrease in energy creates a binding force between the atoms. For example, carbon has four valence electrons grouped in two closely spaced energy levels around the 1 rydberg level (1 rydberg $\sim$ 13·6 eV). It can also accept four more electrons to fill these outer energy states. If four hydrogen atoms are brought close to the carbon atom so that they can share their valence electrons, then all the valence-electron quantum states are filled on a shared basis: two to each hydrogen atom and eight to the carbon atom. This completion of the states leads to a lower energy than in the separated condition and so one obtains the compound methane (CH_4). This shared bond is referred to as the valence bond or *covalent bond*.

The inert gases (helium, neon, argon etc.) demonstrate well the increase in stability when the appropriate energy levels are filled. In these gases the electrons fill up certain groupings of the quantum states of energies, the next higher energies being reasonably well above the full levels. This structure prevents valence electrons being shared with, or accepted by, other atoms, and so the gases (marked with an asterisk in Fig. 2.2) are chemically inert. Now, although it has its outermost energy levels full with 2 electrons, the element calcium is not inert because of the 10 empty quantum states just below the full pair. Similarly nickel is not inert because of its partially full states below its full outermost states. The so-called *transition elements* (atomic numbers 21 to 30 in Fig. 2.2, and others with higher atomic numbers not shown) have two energy levels close together and are characterized by a variable valence. Thus quantum mechanics and the structure of the atomic energy levels explain much about the regular repetition of chemical behaviour as the atomic number changes.

To return to carbon, this element can combine with itself by sharing four valence electrons with four surrounding carbon atoms in an endless array. The atoms around any one atom are centred at the corners of a regular tetrahedron: the tetrahedral bond (Fig. 2.3a). This creates the *diamond* crystal structure. It can be shown that this structure consists of two sets of atoms, each set composed of carbon atoms at the corners and face centres of a cube (face-centred cubic structure, Fig. 2.3b). The two sets are interwoven and joined by the tetrahedral bonds (only one such bond is shown in Fig. 2.3c). Germanium and silicon are two more elements with four valence electrons which can crystallize in this diamond structure. The shared valence electron bond then leads to the nomenclature of *valence crystal* for this type of

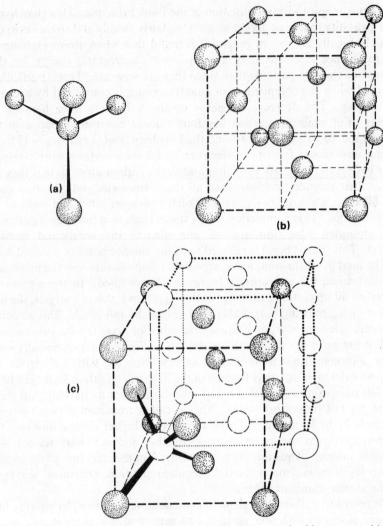

Fig. 2.3 (a) The tetrahedral bond (schematic). (b) The face-centred cubic structure. (c) Interwoven FCC structure. The shaded atoms are A site atoms, the unshaded ones are B site atoms. When A and B atoms are the same, the structure is the diamond; when A and B atoms are different, the structure is called the zinc blende. The tetrahedral bond shows how the atoms of A and B sites are linked. Cube edges define ⟨100⟩ axes; tetrahedral bond defines ⟨111⟩ axes.

crystal. It must be noted that the valence bond does not necessarily create a crystalline structure. Amorphous carbon (graphite) and amorphous silicon are bound by valence bonds, but without the precision that gives the long range order, even though the general tetrahedral arrangement of binding appears to be preserved locally around each atom.

2.3 Ionic or polar crystals

A form of binding force that is conceptually simpler than the valence bond comes from the direct electrostatic attraction between positive and negative charges. In chemical interactions the valence electrons can be exchanged as an alternative to being shared, the exchange occurring in such a manner as to complete the quantum states in certain energy shells. One can then find that the individual atoms are no longer neutral, having either an excess or deficit of electrons. The atoms are said to be *ionized*, and the resulting crystal structures are called *ionic* or *polar*.† An example of a polar crystal is gallium arsenide. This is one of the most studied III-V materials formed between elements from group III (three valence electrons) and group V (five valence electrons) of the periodic table. If the valence electrons transfer from the gallium atom to the arsenic, both will then have a complete shell of eight electrons for their outermost energy (Fig. 2.2) but the gallium atom, having lost three electrons, will have a triple electronic positive charge while the arsenic atom will have a triple negative charge. This would lead to a strong *ionic attraction* between the atoms. However in GaAs the binding forces are not straightforward. It is also possible to share the valence electrons so as to complete the outer energy shells as in valence binding. The actual binding mechanism is then believed to be a mixture of ionic and valence binding, but with the former dominating.

The crystal structure of GaAs is similar to that for C, Si and Ge (Fig. 2.3c) but now the A sites are occupied by Ga atoms while the B sites are taken by As atoms. The crystal structure is commonly referred to as the zinc blende or ZnS structure. Once again the tetrahedral form of binding is preserved, though the detailed mechanism is different. Note also that in the zinc blende structure the central atom of the tetrahedron is always different from the outer atoms.

The classification of different crystal structures[2.15] is beyond the scope of this book but it is of interest to note that the tetrahedral bond can lead to a hexagonal structure. For example, the material cadmium sulphide is a strongly ionic crystal with six valence electrons from the sulphur potentially joining the two valence electrons on the cadmium atom. In general, when like atoms are highly attracted together it is found that one of the structures that they can form is the face-centred cubic while another is the close-packed hexagonal (Fig. 2.4a). The difference in the types of atom prevents the formation of a close-packed hexagonal structure for the CdS crystal, but instead it forms into a closely related hexagonal structure called the *wurtzite crystal form* (Fig. 2.4b). The material CdS is a semiconductor and is also *piezoelectric*.

† The term 'polar' is short for 'heteropolar'; valence binding is also referred to as homopolar binding.

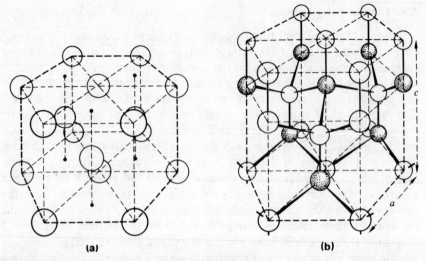

Fig. 2.4 Hexagonal crystals. (*a*) Close-packed hexagonal. (*b*) Wurtzite showing tetrahedral binding.

This means that when it is subjected to fields (especially along the axis labelled *c* in (Fig. 2.4*b*)) the ionic charges on each of the atoms create a force on the crystal which distends the crystal. Equally if the crystal is subjected to pressure, or extension, then the change in the polarizing electric field between the atoms can be detected. In recent years the piezoelectric effect has been used[2.20–2.22] to interact with mobile charge carriers to make novel types of amplifier and filters. CdS has been a very useful material for many basic studies of these effects.

2.4 Metals

In discussing both the valence and the ionic crystals it was assumed that all the electrons of any one atom remained closely associated with the parent atom. Indeed in such crystals the removal of the valence electrons would remove the binding mechanism. The strong localization of the electrons means, in effect, that there are no electrons free to move about and conduct electricity. In a metal the binding mechanism for the atoms is quite different. For example, in copper, silver or gold one finds one outer electron per atom, with all the other electrons forming closed energy shells. It is then found that a crystalline solid forms in which one electron is contributed per atom but this electron is free to move over the whole crystal. The negative electrons form a *jellium* which acts as a bonding fluid holding all the positively charged metallic ions together. The three metals just mentioned crystallize in the same structure, namely the face-centred cubic (Fig. 2.3*b*). This is a structure one obtains from packing spheres as close to each other as possible. (The reader may amuse himself by stacking ping-pong balls together in a tight pack. There are two possible close-packed structures: the face-centred cubic structure of Fig. 2.3b and the hexagonal structure of Fig. 2.4a.)

The lack of direction in the binding given by the jellium of charge, combined with the uniformity of the binding forces, makes these metals readily deformable. Gold is particularly noted for its malleability and ductility and is widely used as a contact material in electronic devices, allowing strain to be taken up in the malleable metal as well as making excellent electrical contact.

Not all metals utilize the electron jellium as the only bond. Nickel, for example, has a strong valence interaction between its atoms, utilizing the electrons from its second highest energy level. Not all the valence electrons will then contribute to the conduction process in nickel.

2.5   Energy bands

When considering the quantum states for electrons in a solid, one should remember that the electron is a wave, so that its 'position' is not as closely associated with its parent atom as the classical theory suggests. The electrons throughout the solid can interact, altering the energy of the local states around the atoms forming the solids. This process mainly effects the outer electrons. As the electrical engineer knows, identical resonant circuits which are coupled together have their resonant frequencies split. Analogously the identical energy levels in neighbouring atoms have their energies separated as the electric fields from one atom interact with the electrons in the other atom. Figure 2.5 shows a standard schematic diagram of what happens in a solid. The diagram indicates 6 atoms, with one quantum state per atom at energy E_0 and 2 states each at energy E_1, being brought closer together. As the atoms move closer the energies split and the groupings of energies may be altered, but the total number of quantum states remains the same as in the isolated atoms. The extreme energies of the groups do not depend on the *number* of atoms but on the *spacing*. Thus a large number of atoms at a given spacing between atoms packs so many quantum states into a given group of energies that it is better to think of a continuum of permitted energy levels—an energy band. This concept is a vital one in classifying materials into insulators, conductors or semiconductors.

Consider first of all the insulators. These are often valence or ionic crystals. In classical terms, one says that the valence electrons are bound to their parent atoms and are not free to gain energy from an applied electric field. In terms of the energy-band theory the equivalent statement is that the energy bands have all their quantum states filled with electrons. The connection is made clear by the Pauli exclusion principle which limits the occupation of any quantum state to one electron. Thus in a full band it will be impossible

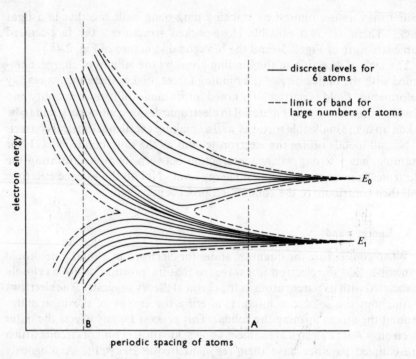

periodic spacing of atoms

Fig. 2.5 Energy band formation. A schematic diagram indicating how two discrete energy levels broaden into bands as the atoms are brought together. If there are six atoms each with 1 state in the upper energy and 2 states in the lower energy, there must always be 18 states as the atoms are brought together, though the groupings of these states may change. If level E_1 were full but E_0 empty, then for a separation of A the material is an insulator, but at B the material would be a conductor. (See, for example, W. Shockley, reference 6.2.)

for any electron to increase its kinetic energy without some equal decrease in another electron's energy. When an electric field is applied it is therefore also impossible for a *net* movement of charge to occur because this would slightly increase the average kinetic energy of all the electrons. Conduction of electricity always implies a net movement of charge and so, in a full energy band, conduction must be impossible. Only if the field was excessively strong, forcing the electrons into the next higher band, would a net motion of electrons be permitted; but this is not the effect for normal conduction. Indeed the very short free paths (~ 10 nm) available to valence electrons for gaining energy from an applied electric field before losing it again in collisions implies that field strengths around 10^8 V/m are required to allow such a jump of around one electronvolt in energy. The band of energies occupied by the valence electrons is the *valence band* and, in the insulator, has just the right number of quantum states to accommodate all the valence electrons. This leads to the band structure of Fig. 2.6a. The inner electrons, those

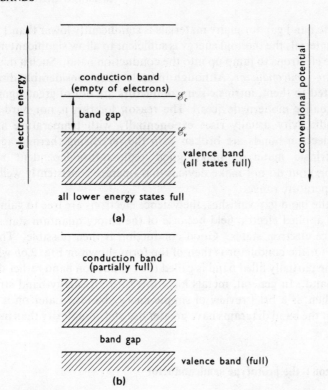

Fig. 2.6 (a) The insulator—if $kT/e \ll (\mathscr{E}_c - \mathscr{E}_v)$, or the intrinsic semiconductor—if $kT/e \sim (\mathscr{E}_c - \mathscr{E}_v)$. (b) The conductor.

even more tightly bound to the parent atom, also completely fill energy bands of lower energy; but they are of little concern here and are rarely shown in band diagrams.

Thermal vibrations can impart energy to the valence electrons, and if the energy is sufficient a few of these electrons will transfer into the next higher band—the *conduction band*. Electrons in this band are free to gain energy from an applied electric field because of the near continuum of empty states above the electrons. Such electrons can then readily contribute to conduction. It follows that the energy gap between the top of the valence band and the bottom of the conduction band in good insulator must be very much larger than any energy that thermal vibrations can impart to the valence electrons. Typically, this implies an energy gap of several electronvolts and certainly in excess of one electronvolt. The reader must be aware of the classical significance of the band gap, namely that it gives the minimum energy which is required to liberate a valence electron bound to its parent atom. It is useful to keep in mind both the electron-band and the bound-electron concepts.

Now the band gap for many materials is significantly lower than 1 eV and in such materials the thermal energy is sufficient to allow significant numbers of valence electrons to jump up into the conduction band. Such a material is an *intrinsic semiconductor*. Although in early texts considerable discussion was devoted to them, intrinsic semiconductors are not of great significance in high-quality modern devices.† The reason for this is not hard to see; their conductivity usually rises exponentially with temperature as more valence electron bonds are broken with the increasing thermal agitation. Thus intrinsic materials make good temperature-dependent resistors (*thermistors*) but do not make devices that operate consistently well over a wide temperature range.

When the band gap vanishes, the valence electrons are free to gain energy from any applied electric field because of the empty quantum states above the valence electron states. Good conduction is then possible. The band structure for the conductor is then of the form shown in Fig. 2.6*b* where by custom the partially filled band is called the conduction band rather than the valence band. In general, metals have this form of energy-band structure.

This, then, is a brief review of simple energy bands. Later on it will be found that the band diagrams have greater use and complexity than indicated here.

2.6   Silicon : the prototype semiconductor

The simple energy-band diagram for a valence crystal like silicon is given by Fig. 2.6*a* if the band gap is taken as 1·1 eV. The precise value of the gap varies slightly with temperature, though this effect will be ignored here. At normal room temperatures the value of the band gap is large enough to prevent all but an insignificant number of valence electrons being thermally excited across the gap. Thus if silicon was made only in its pure state then it would be a good insulator and of little interest.

The objective in semiconductor technology is to make a material of controllable electrical conductivity. This is achieved through the addition of a small number of impurity atoms (*substitutional* impurities) which fit into the crystal lattice in place of the Si atoms but have a different valence. For example, phosphorus which has 5 valence electrons (and an electronic structure that is otherwise similar to silicon's), can substitute for the silicon atoms in the crystal array. However each P atom has an excess electron which does not fit into the tetrahedral bond. This excess electron can be readily detached with a low *ionization* energy or binding energy. This ionization energy can be estimated from the Bohr theory for the hydrogen

† The term intrinsic really means that no impurities, which cause conduction, have been added. In wide band gap semiconductors such intrinsic material is then an insulator.

atom because, in effect, the single excess electron is bound to a single un-neutralized proton in the impurity atom. The effective radius for the orbit of this electron is increased by the high dielectric constant for the material, so that it encompasses several atoms and hydrogen-like binding energy is then reduced to the order of $13\cdot6/\varepsilon_r^2$ eV, slightly less than $0\cdot1$ eV for Si. This heuristic argument gives at least the correct order of the ionization energy ($0\cdot044$ being the measured value for P in Si). At room temperature this binding energy is so small that the bond between the excess electron and parent impurity is nearly always broken and the electron is free to move around the crystal. In the terms of the band structure, the impurity *donates* an electron to the conduction band. Figure 2.7 then shows the simple hand

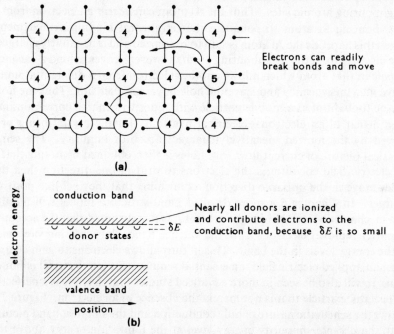

Fig. 2.7 The n-type semiconductor. (*a*) Action of a group V element in a group IV array (schematic). (*b*) Simple band diagram for donor impurities modifying the 'insulator' structure.

diagram for these *donor* impurities, with the quantum state of the donor impurity lying just below the conduction band by the amount of the ioniza-tion energy. The material is kept electrically neutral because for every electron donated to the conduction band there is an ionized, positively charged, donor atom. This then gives the *n-type* semiconductor, and in most of our work the number of conduction electrons in an n-type material is virtually the same as the number of donor impurities. All the group V elements (in particular phosphorus, arsenic and antimony) form donors in Si, though the ionization energies vary from 0·039 eV with Sb to 0·069 eV with Bi.

2.7 Holes

An alternative type of substitutional impurity is an element with three valence electrons, aluminium for example. If this substitutes at the occasional site in the Si lattice then there is a deficit of an electron in the tetrahedral binding arrangement although the crystal is still electrically neutral. It is helpful to remember that electrons are waves that have an influence on neighbouring atomic sites. Thus the Al atom can *accept* an electron from a neighbouring Si atom to complete the tetrahedral binding arrangement. When this happens the Al atom is said to be ionized and is negatively charged. The electron from the neighbouring atom, however, leaves behind a vacancy or *hole* in that atom which allows another electron from a different atom to move into this vacancy and leave the hole in yet another site. Thus the hole can be thought of as a mobile particle with a positive charge corresponding to a deficit of an electron—electrical neutrality for the crystal being preserved by the ionized, negatively-charged, acceptor impurity. The semi-classical picture of current flow with holes is then obtained by noting that if an electric field encourages the electrons to drift in one direction then the holes move in the opposite direction, confirming that they act like positive charges. In Chapter 6 a more detailed band-structure model will modify this mechanistic view, but for the present it is merely noted that the acceptor impurity removes electrons from the valence band, thus creating vacancies in the energy levels in the band. This in turn allows electrons to gain energy from an applied electric field, an essential requirement for electrical conduction. It will also be seen in more advanced studies that the hole is a perfectly respectable particle that is not merely the absence of an electron. Figure 2.8 shows the schematic nature of hole conduction and the simplest band picture with the *acceptor* impurity energy level at the ionization energy above the valence band. When the Bohr theory is applied, *mutatis mutandis*, to holes and ionized acceptors, one would expect similar ionization energies for them as for donor impurities (for Al the ionization energy is experimentally about 0·057 eV). The acceptor impurity creates the *p-type* semiconductor and again, at normal temperatures, the number of impurities determines the number of positive charge carriers in the form of holes.

Almost all group III elements except nitrogen are useful acceptor impurities. The ionization energies range from 0·045 eV for boron to 0·26 eV for tellurium.

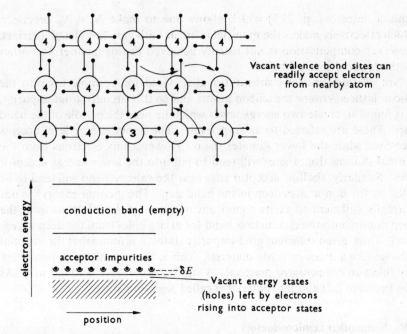

Vacant valence bond sites can
readily accept electron
from nearby atom

conduction band (empty)

acceptor impurities
●●●●●●●●●●----δE
////////////// Vacant energy states
(holes) left by electrons
rising into acceptor states

position

electron energy

Fig. 2.8 The p-type semiconductor. (a) Action of a group III elements in a group IV array. (b) Simple band diagram for acceptor impurities.

Compensation and deep impurities

The feature of fixing conduction properties by the impurity concentration is an important one. Materials with this feature are termed *extrinsic* semiconductors. In extrinsic semiconductors one type of charge carrier is usually greatly in the majority and the minority carrier may be unimportant. In intrinsic semiconductors every electron which is thermally agitated into the conduction band leaves behind a hole in the valence band; thus both types of charge carrier are significant in contributing to conduction in the intrinsic material. One may, then, briefly suppose that an extrinsic semiconductor which was doped with both donors and acceptors could exhibit large amounts of both hole and electron conduction. A sketch of an energy diagram will suggest, however, that the electrons find the lowest energy states that are vacant. Thus for N_d donors and N_a acceptors, with $N_d > N_a$, N_a electrons will fall into the acceptor sites from the donor levels leaving $N_d - N_a$ electrons available for ionization to the conduction band. The electrons in the acceptors will be at too low an energy for thermal stimulation into the conduction band. The material is then n-type in character. Similarly, if $N_a > N_d$, there will be approximately $N_a - N_d$ holes in the valence band. This is simple *compensation* and indeed there is one clever process (see lithium detectors, p. 223) which allows one to make $N_a = N_d$ precisely, which effectively makes the material intrinsic with very few charge carriers. However compensation is not usually achieved in this manner and other methods are discussed below.

Not all impurities act substitutionally. Gold for example fits into the silicon lattice *between* the silicon atoms and so it is an *interstitial* impurity. It is found to create two energy levels which lie near the middle of the band gap. These are referred to as *deep* levels. The upper energy level accepts electrons while the lower donates them. However any electrons from the normal shallow donor sites will tend to fall into the lower-energy acceptor sites. Similarly, shallow acceptor sites near the valence band will tend to be filled by the donor sites deep in the band gap. The thermal energy† is not normally sufficient to excite significant numbers of the electrons from the deep donors into the conduction band (or excite holes from the deep acceptors). Thus, given sufficient gold-impurity states, it is found that the material behaves like a quasi-intrinsic material. This is the common situation when one refers to *compensated* material. A similar occurrence is found in GaAs (see Problem 2.2) and the material is called *semi-insulating*.

2.8   Some other semiconductors

For any semiconductor to be useful in well-engineered devices it must have a well-developed technology of production, purification and processing. Germanium was at one stage the semiconductor with the best set of these three Ps. However, it has a band-gap energy of only 0·65 eV, so that when it is hot, significant numbers of valence electrons can be agitated thermally into the conduction band: its conductivity is temperature-dependent. Silicon is much better in many respects and is currently the material with the best set of three Ps.

More recently, the technology of gallium arsenide has reached a very high standard, especially with epitaxial growth (see section 2.9). This material has a band gap of about 1·4 eV and in its *pure* state is a good insulator. It is possible to make GaAs which comes out in the 10^6 Ωm resistivity range, although this is usually compensated material with perhaps oxygen or chromium acting as the deep compensating levels. Like silicon, gallium arsenide can have substitutional impurities which will either donate or accept an electron. For example zinc, with two valence electrons, will substitute for a gallium atom with three valence electrons, and so zinc will act like an acceptor or p-type impurity. Sulphur has six valence electrons and will substitute for the arsenic atom, giving a donor or n-type impurity. In both these

† kT is usually much less than half the band gap energy.

examples the impurities are readily ionized at room temperature and so give rise to one electron or hole per impurity atom, at least for low impurity concentrations. The behaviour of the elements from group IV when used as impurities in GaAs is interesting. Germanium, for example, at low concentrations can substitute for the Ga atom and so act as a donor. However, at high concentrations it will start to substitute for the As atom, leading to a more p-type impurity and preventing further increase in the n-type impurity. This type of behaviour is said to be *amphoteric*. Tin is a similar amphoteric impurity.

Cadmium sulphide has already been mentioned as a piezoelectric material but it is also a light-sensitive material. It has a band gap around 2·4 eV, so that in its pure state it is a good insulator. However the quantum unit of electromagnetic radiation, the photon, has sufficient energy, when the radiation is in the visible part of the spectrum, to give a valence electron enough energy to jump from the valence band into the conduction band. This is one of the main reasons for the sensitivity of CdS to light.

Indium antimonide is an interesting material on account of its very small band gap: only 0·16 eV. This makes it an intrinsic semiconductor with its conductivity very temperature-dependent. The electrons in the conduction band, however, can move extraordinarily fast for small electric fields. In the jargon to be developed later, it has a high *mobility*. It also has the feature that the electrons move in the conduction band as if they had only $\frac{1}{75}$ of the free electron mass. This is a puzzling feature until one considers effective mass in the more detailed discussions about band structure (see Chapter 6).

Semiconductors that are made from combinations of materials are steadily coming into use as technology improves. One finds that one can change certain important features of the band structure of the materials (e.g. band gap) by changing the stoichiometry of the different materials. For example GaAs and AlAs can be mixed to produce $Ga_xAl_{1-x}As$, which is a semiconductor whose band gap can be varied from 1·4 eV ($x = 1$) to 2·4 eV ($x = 0$). There is always a limit however to the usefulness of these techniques because of the expense of bringing any one material up to the required standard of production, purification and processing for commercial use.

2.9 Production of pure material

We have seen that it is the impurities that control the numbers and type of charge carriers in an extrinsic semiconductor, and so it will be useful to estimate the impurity concentration required for useful conduction. To do this in detail will need much work but one can begin by noting that copper, with one electron per atom contributing to conduction, has a resistance around 10^{-5} ohms for a cubic millimetre. Resistances that are used in electronic circuits will require up to 9 or 10 orders of magnitude higher value.

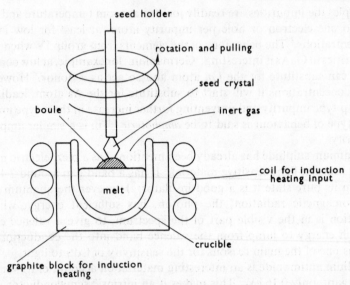

Fig. 2.9 The Czochralski method.

Thus one might expect only one atom in every 10^9 to contribute a conduction electron (or hole). Clearly this range must vary by several orders of magnitude but it is obvious that the problem is not one of adding impurities so much as one of excluding unwanted impurities down to a level of 1 part in 10^{10}. Great care has to be taken to produce single crystals of this quality, and they are expensive.

Any material that will crystallize can be crystallized by slow cooling from a molten mass, or by cooling a supersaturated solution of the material. The classic home experiment is that of cooling a supersaturated solution of copper sulphate in water; crystal platelets will readily form as the liquor cools. A much larger crystal can be grown if a seed crystal of copper sulphate is suspended in the solution as it cools; the growth is then onto the seed crystal. One of the best proven techniques for the growth of large semiconductor crystals is the Czochralski method (Fig. 2.9).[2.1,2.16] The essential features are the maintenance of the molten mass of raw material (the *melt*) at a controlled temperature that is close to the solidifying temperature. Into this melt is inserted a small seed crystal to which the atoms from the melt will attach themselves, building upon the pattern of atoms formed by the seed. Thus the seed will determine the orientation of the crystal growth. To ensure uniformity of growth, the seed is slowly rotated and pulled out of the crystal at just the correct rate (usually a millimetre or so per hour). The resulting crystal is called a *boule* and may be several centimetres in diameter and a good fraction of a metre in length. Some materials, for example

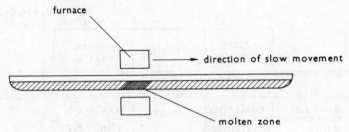

furnace

direction of slow movement

molten zone

Fig. 2.10   Zone refining.

gallium arsenide, must be used very carefully. Arsenic is likely to boil off the melt, which has to be kept under pressure of an inert gas to prevent this; one may also have a layer of suitable molten glass over the melt as a further protection. Material grown by this process is referred to as *bulk grown*.

Such a bulk growth is only the first stage in obtaining pure material. Fortunately a pure crystal is in a lower state of energy than an impure crystal because of its energy of cohesion. This means that as a crystal grows it has a tendency to reject both impurities and defects. Every time that a material recrystallizes there is thus a tendency for the impurities to be left behind in the molten part and for the pure material to crystallize out. The fraction of impurities remaining at each crystallization tends to be a constant, called the *distribution coefficient* for the impurity in question. This effect can be used to advantage in *zone refining* (Fig. 2.10), a process[2.18,1.4] in which the crystal is made molten in a local zone and the zone is slowly moved down the crystal, driving the impurities to one end. Zone refining leaves the material with a residual impurity concentration that is at a much lower level but is perforce graded; this limits the usefulness of the technique. Impurities can be added to the melt to make it p-type or n-type.

It must not be supposed that the art of crystal growth is really as simple as these few paragraphs might suggest. Great care has to be taken to ensure uniformity of the growth conditions so that defects and dislocations do not grow into the crystal. Reference 2.17 gives a readable account of the types of disorder that can occur in a crystal. If the required purity is to be achieved, starting chemicals have to be very pure and growth conditions very clean.

Epitaxial growth

The use of very pure starting materials and the segregation of impurities during crystal growth have made possible an effective method of growing extremely pure, thin layers of semiconducting material on top of seed crystals or *substrates*. The thin skin is called an *epilayer*, giving the name *epitaxial growth*, coined from Greek (*epi-*, prefix for upon; *taxis*, arrangement). In many practical devices the active regions of material need to be only

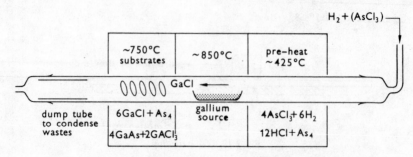

Fig. 2.11   GaAs vapour-phase epitaxy (schematic).

micrometres (*microns*) thick but will require electrical contacts to both sides. An epitaxial growth allows the substrate to act as one of the electrical contacts provided that the substrate is heavily doped with the appropriate impurity to raise its conductivity. The substrate can then be thick enough to give the very thin layer a backing for strength and yet contribute only a low contact resistance. The epilayer can have its impurity doping determined quite independently of the substrate, and indeed will be doped for optimum performance of the device being manufactured.

Some of the most successful techniques for epitaxial growth have been achieved by allowing a vapour containing the semiconductor atoms to condense around slices of the substrate crystal. One such scheme is shown for GaAs in Fig. 2.11. The substrate slices will be cut from a bulk-grown crystal and be typically a few centimetres in diameter and a millimetre or so thick. The system of growth is continuous once it has been started, and the rate of growth depends upon the flow rates of the gases transporting the impurities and the temperature of the growing zone. In general, the thickness is proportional to the time of growth. Specific impurities can be added into the gas stream (e.g. the hydrogen shown in Fig. 2.11 could be passed over a boat of sulphur at low temperature; the sulphur vapour would add a small amount of n-type impurity). Care must be taken not to allow any impurities to contaminate the system permanently. There are other techniques of growth;[2.19] *liquid-phase epitaxy* for example has been found to be useful in the manufacture of materials for light emitting diodes.[9.8] In this process the epilayer is condensed from a slowly cooling melt containing the semiconductor material, in liquid form, and any required impurities.

An important extension of epitaxial techniques is that of growth in selected areas using *oxide masking* (Fig. 2.12). The ideas associated with this will have a very wide application to many devices and so it is convenient to introduce these ideas here. Thin layers ($\sim 10^{-6}$ m thick) of silicon dioxide (SiO_2) can be grown onto silicon by heating the material in a moist oxygen atmosphere, or onto other materials by the thermal decomposition of silane gas (SiH_4:

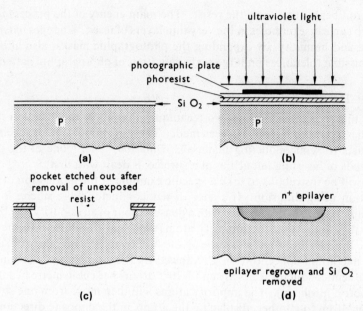

Fig. 2.12 Masked epitaxy technique. (*a*) A layer of SiO$_2$ is grown, the thin layer of photoresist is placed over the slice (*b*) and exposed in certain areas, defined by a photographic plate, to ultraviolet light. The unexposed resist is washed away, the pocket of silicon and dioxide etched out (*c*), the exposed resist removed, and the slice placed in an epitaxial reactor. Finally the SiO$_2$ may be removed (*d*). Photoresist by itself will not protect at high temperatures.

an explosive if used carelessly with oxygen). Such layers of SiO$_2$ will protect the material from attack by certain chemicals and in particular will prevent epitaxial growth from adhering to the main slice. However, the oxide can be etched away by chemicals to expose the semiconductor underneath. The etching can be done in selected areas by the process of *photolithography* (Kodak give very full details[2.13]). In essence, after the deposition of the SiO$_2$ the surface of the material is covered with *photoresist*, a rubbery material that hardens on exposure to ultraviolet light and will, in its hardened areas, resist chemical etchants (Fig. 2.12*a*, *b*). In the unexposed areas the resist can be washed away by solvents, and etchants can then attack the material beneath. The hardened resist can be removed by more prolonged soaking in solvents and the slice processed once more with the SiO$_2$ now protecting the areas against the epitaxial growth adhering to the main crystal. The photolithographic process clearly has wider applications, as we shall see, and it should be noted that resists which are of the opposite character (protect in unexposed areas) can be obtained. The process can be made very accurate; tolerances of around a micrometre can be obtained with the use of light and sub-micrometre tolerances have been obtained using special resists and electron beams to *expose* the resist. The main enemy of the process is dirt, which can cause pinholes in the very thin layers of resist. Changes of temperature and humidity, by expanding the photographic masks, also affect the permissible tolerances on large-scale definition of photographic patterns.

2.10   The addition of impurities

In this section we outline two techniques that allow the insertion of impurities *after* the material has been made. These are important for specialized contacts to semiconducting materials. *Diffusion*,[2.10,2.11] one of the oldest methods of inserting impurities at a surface, is dealt with first.

It will be instructive to take a specific example of the fabrication of a p-n junction using diffusion. If a slice of semiconductor (take silicon for this example) is placed in a furnace with a dense vapour of an impurity (say boron) then it is found that the impurity atoms will tend to diffuse into the solid slice. Initially the atoms will settle on the surface and dissolve into the surface at a maximum concentration C_s allowed by the solubility of the impurity in the semiconductor (C_s is around $3 \cdot 10^{26}$ atoms per cubic metre at 1100 °C for boron in silicon). The impurity atoms will then move from one site of a silicon atom to another, displacing the atoms in the opposite direction. At a given concentration the movement is quite random, with as many atoms moving one way as the other. Thus there is only a *net* movement of the impurities from the high concentrations into the lower concentrations, that is to say towards the middle of the material. This movement is described by Fick's law, which gives the relationship between the concentration $C(x)$, at the point x, with the flux of impurities (number crossing unit area/time) $F(x)$ by

$$F(x) = -D \, \partial C / \partial x \qquad\qquad 2.1$$

D is the *diffusion constant*; this typically varies with temperature as $D = D_0 \exp(-eV_a/kT)$. The significance of eV_a is that it represents an energy (in electronvolts) that is required to displace a crystal atom by the impurity atom. For boron in silicon this energy appears to be close to 5 eV so that, even though $D_0 \sim 10^8$ m^2/h, it still takes around ten hours for boron to diffuse a few micrometres into silicon at 1100 °C. At normal room temperatures the diffusion process is quite negligible. It is shown in Appendix 1 how to solve eqn 2.1 to give the impurity profile $C(x, t) = C_s$ erfc $x/(4Dt)^{1/2}$ (Fig. 2.13). It is common to deposit the impurities for a short interval t_0, then 'turn off' the supply and continue the diffusion process to *drive in* the impurities. The number of impurities inserted in the pre-deposition step is $Q = (4Dt_0/\pi)^{1/2} C_s$ per unit area and these are then redistributed. The subsequent profile follows the law $C(x, t) = [Q/(\pi Dt)^{1/2}] \exp(-x^2/4Dt)$ (Fig. 2.13). Note that the surface concentration now falls with time during this drive-in procedure.

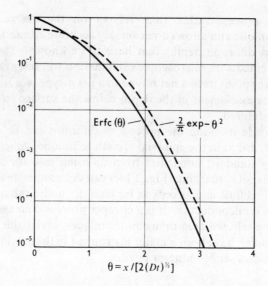

Fig. 2.13 Diffusion profiles. Deposition gives C_s erfc (θ) for time $t = t_0$. Drive-in gives $(2/\pi)C_s(t_0/t)^{1/2} \exp - \theta^2$ for time t of diffusion $\gg t_0$.

Figure 2.14a shows a schematic system for diffusion of boron. The impurity is carried in a vapour by a gas stream similar to the epitaxial growth technique. If phosphorous had been required then $POCl_3$ could have been used instead of BCl_3. The drive-in diffusion might be done by turning off the gas supply to the impurity. Figure 2.14b shows a technique for calibrating

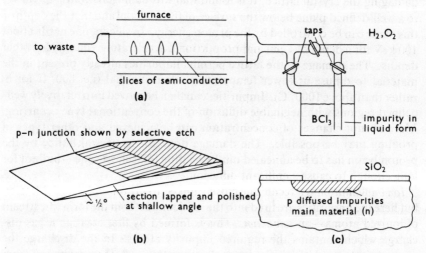

Fig. 2.14 Diffusion. (a) Schematic diagram of furnace system. (b) Angle-lap technique for measurement of diffusion depth. (c) Selective area diffusion by oxide masking.

the diffusion process rather than relying on the theory. This is the *angle-lap* technique and allows a reasonably accurate estimation to be made of the shallow diffusion depths that have to be known. Special chemical etchants combined with microscopic examination will reveal the position of the change-over point from a net p-type to a net n-type impurity and so help in the accurate assessment of the depth below the surface to which the impurities have diffused.

By using oxide masking the process of diffusion can be carried out in selected areas, just as in the epitaxial growth technique. Silicon dioxide will prevent many standard impurities from diffusing into the semiconductor beneath. Thus SiO_2 masking (Fig. 2.14c) can determine areas for diffusion. In general, the diffusion proceeds as far laterally under the oxide as it goes down into the semiconductor. It is a cheaper process than epitaxial growth, and indeed the selected area diffusion techniques, under the trade name of the *Planar process*, have been a major instrument in the advance of low-cost, high-performance, semiconductor devices.

Ion implantation[2.12]

When impurities are inserted into a layer of semiconductor by diffusion, the *profile* (the change of impurity with depth) is limited to a Gaussian or error function. It is difficult to make extremely abrupt changes of concentration; while epitaxial techniques help, there are newer methods which offer the possibility of choosing the impurity profile. One such method is *proton-enhanced diffusion*, where energetic H^+ ions are fired into the semiconductor, damaging the crystal lattice. It is found that this damage is confined closely to a well-defined plane below the surface of the material and that the depth of this plane can be controlled by the proton energy. In silicon, one needs about 100 keV for a depth of 1 μm and proportional accelerating voltages for other depths. The damage to the lattice permits impurities already present in the material to diffuse at lower temperatures than is usual (i.e. 800 °C for Si rather than above 1000 °C). Impurities can then be moved into relatively well-defined regions with negligible diffusion of the conventional type occurring. More abrupt changes of concentration can be achieved and other types of profiling may be possible. The damage done to the crystal lattice by the proton beam has to be annealed out by heating for a short time—but not for long enough to cause significant diffusion.

Ion implantation is a technique closely linked to proton-enhanced diffusion but here the impurities are fired into the semiconductor in the form of a stream of ionized atoms—an *ion beam*. This is formed by first creating a gas discharge which contains the required impurity atoms. In the discharge the atoms are stripped of their valence electrons (*ionized*), then drawn off from the gas discharge by suitable apertures and focusing electrodes to form an

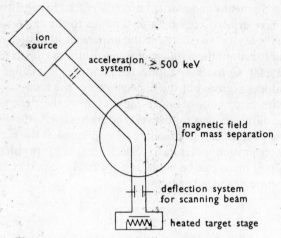

Fig. 2.15 Schematic layout of ion implantation unit.

ion beam. If a gas which contains the desired impurity atoms cannot be found, one uses some inert gas in the discharge. The energetic ions in this gas can then knock out, or *sputter*, the required impurity atoms from some solid material which does contain them. The sputtered ions can be drawn off along with the other ions. The required impurity ions are selected by passing the beam through a magnetic field which gives different curvatures in the trajectory of particles with different (charge/mass) ratios. The selected impurities can then pass through an aperture or slit to be focused and accelerated into the semiconductor material. Figure 2.15 shows a schematic diagram of an ion-implantation unit.

The ions forming the beam have sufficient energy to bury themselves in the semiconductor. For amorphous material, a rough guide to the depth of penetration is given by the *projected range* R_p

$$R_p \sim \frac{39}{g} V_0 \frac{M_h + M_i}{M_h + 3M_i} \frac{1}{Z_i^{2/3}} \, \mu\text{m}$$

where V_0 is the accelerating potential in megavolts and g is the material density (g/cm³) with M_h and M_i the mass of the host lattice atom and implanted ion respectively. The atomic number of the inset atom is Z_i. There is of course a spread in this range typically given by an empirical relationship: $\delta R/R \sim 0.7(M_h M_i)^{1/2}/(M_h + M_i)$. Although the depth of penetration into an amorphous material is fairly predictable, the range can be increased by an order of magnitude in a crystalline material by firing the ions along the lines of the crystal axes. The ions are then *channelled* along the crystal structure. Just how far such channelled ions travel depends critically on the crystal orientation.

The ions, when implanted, cease to be ions and become impurity atoms but they are by no means necessarily correctly placed in substitutional sites. They usually have created considerable damage to the crystal structure on their way into the crystal. To activate the impurities in the desired substitutional sites and to reduce the damage one *anneals* the crystal at temperatures of around 400–600 °C, for Si. At these temperatures, negligible diffusion of the impurity atoms occurs, but the dislocations of the atoms and the interstitial impurities can readjust under the action of the thermal agitation. Annealing, like diffusion, may take several hours to achieve the desired result. In summary, impurities can in principle be implanted at the required depth by different acceleration potentials. Special impurity profiles, especially shallow and sharp ones, can be made and are likely to find commercial use in the manufacture of specialized devices.

PROBLEMS

2.1 An electron is orbiting around a proton with a circular orbit of radius R such that the circumference is an integral number of wavelengths given by the free electron wavelength formula $p = h/\lambda$. Verify that the angular momentum is an integral number of h units. By balancing the centrifugal force against the electrostatic attraction, show that the radius of the orbit is limited to values $r = (4\pi\varepsilon_0 n^2 h^2/e^2 m)$ (for the purposes of this calculation one need only assume that the proton is virtually stationary compared with the electron; the small error in this assumption is of little concern here). Hence show that the energy required to remove the electron from the influence of the proton is at most $\{e^4 m/2(4\pi\varepsilon_0)^2 h^2\}$. Hence show that r_0, the orbit radius for the ground state, is of order 0·05 nm and the energy is 13·6 eV.

How is this energy of escape affected if the electron effectively orbited in a material of relative permittivity ε_r? How is the orbit radius affected by such a change and by an effective electron mass that is lighter than the free electron mass?

2.2 Read Chapter 3, in particular the section about Fermi level.

Oxygen impurity in GaAs has a donor level at approximately 0·63 eV above the valence band. Zinc gives a shallow acceptor at 0·024 eV above the valence band, while Si gives a shallow donor at 0·002 eV below the conduction band. The band gap may be taken as 1·4 eV. If there are $10^{20}/m^3$ Si impurities, $10^{22}/m^3$ O impurities, and $10^{21}/m^3$ Zn impurities, show that the Fermi level is approximately 2·4 kT/e above the deep O level. The effective density of states for the conduction band in GaAs is $5 \cdot 10^{23}/m^3$ while for the valence band it is $7 \cdot 10^{24}/m^3$ (*Hint*: most of the detailed values given above do not matter; only their orders of magnitude are important).

2.3 The making of a device requires an epilayer 4 micrometres deep for its active n-type region. The device is to be protected by a micrometre of SiO_2, but for every micrometre of growth of SiO_2, approximately 0·5 micrometre of Si is used up. The high density of n-type impurities in the substrate out-diffuses into the low-density epilayer whenever the epilayer and substrate are raised to diffusion temperatures. By suitable choice of impurities this out-diffusion of impurities has been kept to one-quarter of the in-diffusion process for the p-type top contact. The contact is to be a deep (4 micrometre) diffusion of boron. Bearing in mind that the manufacturer will supply to only a 20 per cent tolerance, what thickness epilayer would you order? What alterations in the process could accommodate the variation in the thickness? Estimate the diffusion time if the surface concentration is kept high and the main epilayer has an impurity content at $10^{21}/m^3$ The diffusion furnace temperature may be taken as 1100 °C. [Answer: 11 micrometres. 9–10 hours.]

General references

2.1 KITTEL, C. *Introduction to Solid State Physics*. 4th edition. Wiley, 1970.
2.2 SOLYMAR, L. and WALSH, D. *Lectures on the Electrical Properties of Materials*. Oxford University Press, 1970.

Special references

Quantum physics and atomic structure

2.3 WEHR, M. R. and RICHARDS, J. A. *Physics of the Atom.* Addison-Wesley, 1960.
2.4 SLATER, J. C. *Atomic Theory of Atomic Structure.* Vol. 1. McGraw-Hill, 1960.
2.5 DICKE, R. H. and WITTKE, J. P. *Introduction to Quantum Mechanics.* Addison-Wesley, 1961.
2.6 WHITE, R. L. *Basic Quantum Mechanics.* McGraw-Hill, 1966.
2.7 SCHIFF, L. I. *Quantum Mechanics.* McGraw-Hill, 1955.
2.8 BORN, M. *Atomic Physics.* 8th edition. Blackie, 1969.
An old favourite with students that is helpful in giving a perspective on many points when read in conjunction with a more modern text.

Semiconductor materials and techniques

2.9 KANE, P. F. and LARRABEE, G. B. *Characterisation of Semiconductor Materials.* McGraw-Hill, 1970.
2.10 GROVE, A. S. *Physics and technology of semiconductor devices.* Wiley, 1967.
2.11 WOLF, H. F. *Semiconductor.* Wiley, 1971.
A useful source of information on silicon technology.
2.12 MAYER, J. W., ERIKSSON, L., and DAVIES, J. A. *Ion implantation in semiconductors.* Academic Press, 1970.
2.13 Eastman Kodak. KODAK PHOTO-RESIST SEMINAR 1968. Vols. 1 and 2.

Crystal structure, growth etc.

2.14 KITTEL, C. *Introduction to Solid State Physics.* 4th edition. Wiley, 1970.
2.15 PHILLIPS, F. C. *An Introduction to Crystallography.* 3rd edition. Wiley, 1963.
2.16 LAWSON, W. D. and NIELSEN, S. *Preparation of Single Crystals.* Butterworths, 1958.
2.17 KITAIGORODSKIY, A. I. *Order and Disorder in the World of Atoms.* Longmans, 1967.
2.18 PFANN, W. G. Principles of zone-melting. *J. Metals,* **4** (1952), 747–753.

Epitaxial growth

2.19 *RCA Review* Vol. 24 (1963), December.
This volume contains several interesting items including a classic paper by H. Nelson on liquid epitaxial growth of GaAs, and the use of silane gas to grow silicon epitaxy, a paper by S. R. Bhola and A. Mayer.

Ultrasonics

2.20 QUATE, C. F. and BLOTEKJAER, K. The coupled modes of acoustic waves and drifting carriers in piezoelectric crystals. *Proc. IEEE,* **52** (1964), 360–377.
2.21 WHITE, R. M. Surface elastic waves. *Proc. IEEE,* **58** (1970), 1238–76.
2.22 MITCHELL, R. F. Acoustic surface wave filters Philips Tech. Rev. **32** (1971) 179–89.

3

Conduction and Equilibrium in Solids

3.1 Introduction

This chapter is concerned with the laws governing the motion of charge carriers in a solid and with the equilibrium conditions that are established for these carriers. The relevance of the different sections may not always be apparent until later chapters where the particular effects are used. It is, however, convenient to place the basic work in a single chapter so as to avoid repetition and dislocation of the discussions on devices. Consequently the reader is advised to give this chapter a quick initial reading, to grasp the central ideas, and then to return to the details as required in the later sections on practical devices.

The starting point should be familiar to most readers: Ohm's law. This states that the rate at which charge flows through a conductor is proportional to the voltage applied across the terminals of the conductor. A simple classical model will be developed to give this law, on the assumption that the conducting material is uniform and electrically neutral. It will then be shown that any accumulations of charge inside such a conductor are rapidly dispersed, with a time constant called the *dielectric relaxation time*, so that one is assured of the validity of the assumption of electrical neutrality in many practical examples. The discussion then leads naturally on to show that time-varying fields can contribute to a current flow just as well as a movement of charge. This contribution is the *displacement current*. All these arguments that apply to conductors can be applied to semiconductors, but in the semiconductor one often finds an additional component of current flow becoming prominent. This additional current flow is caused by a net movement of charge, even in the absence of a field, whenever there is a spatial gradient in the concentration of charge carriers. This is the *diffusion current*, and it is created by the random thermal motion of the charge carriers. This process is not to be confused with the diffusion of impurity atoms, although it is closely related in form; there is no movement of the impurities in the conduction processes considered in this book.

The equilibrium between different materials can either be discussed in terms of the diffusion of charge carriers or discussed using a statistical approach. Statistical methods are powerful, but space allows only a brief treatment to establish the important concept of the Fermi level. This level is a reference energy important in all types of solid. The statistical method using the Fermi level will show that, as well as electrical equilibrium, there has to be an equilibrium between the numbers of holes and electrons in a semiconductor. There are always a few charge carriers, *minority carriers*, of opposite sign to the majority charge carriers determining the basic nature of the material. The chapter is concluded with a short section which shows how equilibrium concentrations are established.

3.2 Ohm's law

Except for the points where quantum mechanics modifies the picture, a mainly classical model of current flow will be used to explain Ohm's law. Initially it will be convenient to suppose that the current is created by the motion of mobile electrons; the motion of holes can be added later. One finds that, as the electrons move inside a crystal, quantum mechanics demands an alteration of the electron's mass from its free-space value to an *effective mass*; this typically lies in the range of $\frac{1}{100}$ to 2 times the free space mass. The manner in which this change arises is left until Chapter 6 and for the moment the reader is asked to accept this effect—perhaps as he might accept the change in apparent mass of a body immersed in a fluid.

In the absence of any electrical field one should have the picture of a completely random motion of the mobile electrons in the solid: no net movement in any one direction. This random movement is characterized by a mean energy per electron that is directly related to the temperature of the crystal. One frequently refers to an 'electron gas' so as to represent this picture. As in a gas, the mobile electrons in the crystal are colliding with each other; they also are colliding with the atoms of the crystal, and with impurities and defects. For electron–atom collisions quantum mechanics modifies the picture quite drastically. It is found, because of the wave motion, that electrons can actually travel freely through a perfectly periodic crystal without any collisions with the lattice; there will be collisions only with those atoms which are displaced from their ideal periodic positions in the crystal lattice. However, even with the *perfect* crystal, atoms are frequently displaced from their ideal positions in the lattice because of the thermal vibrations of the lattice. Such vibrations therefore contribute to the cause of collisions by the charge carriers; the collisions will then contribute to the electrical resistance. In general the thermal vibrations of the lattice increase with temperature and so the electrical resistance increases with temperature.

In the discussion here, the effects of all the collision processes are lumped together into a single parameter τ_m, where τ_m is the average time between collisions which reduce the momentum of a charge carrier to zero. One can be a little more precise about this definition by supposing that each electron in some given volume was labelled, at an instant in time t, with the time t_n that it last had zero momentum in some specified direction $\overrightarrow{OX}$. If there were N electrons then τ_m would be defined by

$$\tau_m = \sum_{n=1}^{N} (t - t_n)/N \qquad 3.1$$

For most materials of interest it will be found that this average value of τ_m is independent both of the time t at which the observation was made and of the specified direction $\overrightarrow{OX}$. Equation 3.1 can be assumed still to hold in the presence of a small electric field $-\mathbf{E}$ in the $\overrightarrow{OX}$ direction. The electrons at the time t_n can be assumed to have some random velocity $\mathbf{u}_n$ at right angles to the direction $\overrightarrow{OX}$ (it is only the momentum in the direction of $\overrightarrow{OX}$ that is zero at t_n) so that at the time t the electrons velocity is given by

$$m^*\mathbf{v}_n = (-e)(-\mathbf{E})(t - t_n) + m^*\mathbf{u}_n \qquad 3.2$$

Averaging over all particles, and remembering that $\mathbf{u}_n$ is random and so averages to zero, one obtains (using eqn 3.1) the average particle velocity $\mathbf{v}$:

$$m^*\mathbf{v} = e\mathbf{E}\tau_m \qquad 3.3$$

or $$\mathbf{v} = \mu\mathbf{E} \quad \text{where} \quad \mu = e\tau_m/m^* \qquad 3.4$$

The parameter μ is called the *drift mobility* of the electrons and has the dimensions of m/s per V/m ($\text{m}^2/\text{V.s}$). Its order of magnitude for many practical semiconductors lies in the range 0·01 to 1 $\text{m}^2/\text{V.s}$. An analogous argument holds for holes, and a mobility can be similarly derived. The reader should contrast the motion here with that of free electrons where the velocity continues to increase in a steady field. It must of course be noted that eqn 3.3 is only valid if the electric field is substantially constant on a time scale comparable with τ_m.† This however will be commonly true for work in this book where τ_m typically ranges from 10^{-12} to 10^{-15} sec. A more advanced discussion of the effects of collisions requires the introduction of the Boltzmann collision equation (see references 3.1, 3.2 and 3.8). The parameter τ_m is then correctly described as the momentum relaxation time (i.e. the time scale required to readjust the momentum by a change of electric field).

Equation 3.4 leads directly to Ohm's law and is itself sometimes loosely referred to as Ohm's law. Consider a block of material, satisfying eqn 3.4,

---

† See reference 2.1 for an elementary treatment, where $\omega\tau$ is greater than unity which is relevant for conduction at optical frequencies in metals.

with a length L and cross-sectional area A containing n_0 electrons per unit volume. A voltage is applied across the length, giving rise to a field $E = -V/L$. The current is then the amount of charge reaching a terminal in unit time and is thus $I = en_0A\mu(V/L)$, with the current opposing the voltage that gave rise to it. Hence, rearranging, one obtains:

$$V = IR \quad \text{where} \quad R = L/A\sigma \quad \text{and} \quad \sigma = en_0\mu \qquad 3.5$$

This is Ohm's Law; R is the resistance, σ is the *conductivity* and $1/\sigma$ is called the *resistivity*. Typical values of resistivity for semiconductor materials range from 10^5 ohm m to 10^{-5} ohm m. When both holes and electrons are present in any numbers, Ohm's law still holds but $I = e(n_0\mu_n + p_0\mu_p)EA$ where p_0 is the hole concentration and the suffices n and p distinguish electron and hole mobility. The holes drift in the opposite direction to the electrons but because the sign of their charge is positive the current is still contributed in the same sense.

3.3 Gauss' theorem, charge continuity and electrical neutrality

One of the first points that should be noted about Ohm's law is that it totally neglects the fields which arise from the charge carriers themselves; any electric field is supposed to be applied by external means. This can only be a valid assumption if the fixed background charges neutralize the mobile charges. In considering how free holes and electrons were formed, it was seen that, in principle, electrical charge neutrality would be ensured because for every ionized donor impurity (positive charge) there was an electron in the conduction band, and equally for every ionized acceptor atom (negative charge) there was a hole in the valence band. The net overall charge is thus zero. This condition is shown here to be a highly stable situation provided that the fixed ionized impurities are uniform and there are sufficient of them. The discussion will require the use of Gauss' theorem, one of the fundamental theorems of electromagnetic theory.[3.9]

Gauss' theorem states that the total electric flux (given by the normal component of the flux density vector $D = \varepsilon_0\varepsilon_r E$ integrated over the surface) out of a volume is equal to the total enclosed charge. The reader must fully understand the one-dimensional form of this theorem, which will be used time and again. In Fig. 3.1a is shown a region of area A and thickness δx bounded by parallel planes. A field $E(x)$ is normal to these planes and in between the planes is a net charge ρ per unit volume. The dielectric constant is assumed uniform, so that Gauss' theorem yields

$$A\varepsilon_0\varepsilon_r[E(x + \delta x) - E(x)] = \rho A\delta x$$

The left-hand term is the net contribution of the electric flux vector $\varepsilon_0\varepsilon_r E$ out from the volume $A\delta x$ (ε_r is the relative dielectric constant); the right-hand

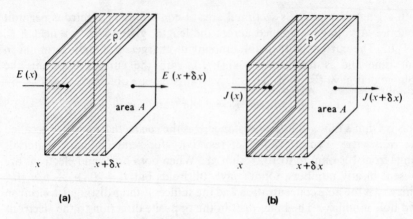

Fig. 3.1 Notation for (a) Gauss' theorem and (b) continuity of charge.

term is the enclosed charge. Now expand $E(x + \delta x)$ to $E(x) + \delta x \dfrac{\partial E}{\partial x}$ where terms of order $(\delta x)^2$ have been omitted and obtain immediately the differential form of Gauss' theorem required here (see Problem 3.1) as

$$\varepsilon_0 \varepsilon_r \frac{\partial E}{\partial x} = \rho \qquad\qquad 3.7$$

The second theorem that will be required is the charge continuity theorem, another fundamental relationship which states that the charge is neither created nor destroyed. This theorem may be developed in a similar manner to eqn 3.7. In Fig. 3.1b is shown a charge ρ trapped between two planes exactly as before but now there is a current of $J(x)$ per unit area (*current density*) flowing into the element of volume $A \, \delta x$ and similarly $J(x + \delta x)$ moving out. The rate of increase of the total charge inside the element of volume must be given by the net rate at which charge is being fed into the volume; consequently one has

$$A \, \delta x \frac{\partial \rho}{\partial t} = A[J(x) - J(x + \delta x)]$$

Expanding the last term by the Taylor expansion as previously, and neglecting terms of order (δx^2), leads to the differential form of the continuity equation

$$\frac{\partial \rho}{\partial t} = - \frac{\partial J}{\partial x} \qquad\qquad 3.8$$

If Ohm's law is now used, the changes of current density δJ are related to the changes of electric field by a conductivity σ_1 so that

$$\delta J = \sigma_1 \, \delta E \qquad \left(\sigma_1 = \rho \, \frac{dv}{dE}\right) \qquad\qquad 3.9$$

Combining eqns 3.7, 3.8 and 3.9 yields

$$\frac{\partial \rho}{\partial t} = -\frac{\sigma_1}{\varepsilon_r \varepsilon_0} \rho \quad \text{or} \quad \rho = \delta P \exp -\sigma_1 t / \varepsilon_r \varepsilon_0 \qquad\qquad 3.10$$

Equation 3.10 therefore shows that any net charge density δP decays with a time constant $\tau_d = \varepsilon_r \varepsilon_0 / \sigma_1$, a characteristic parameter called the dielectric relaxation time. It is found that for many practical semiconductor devices the material used has values of τ_d less than a picosecond. In such examples one is usually well justified in talking about electrical neutrality. In a metal, the dielectric relaxation time is so short that one does not consider the metal to be anything but electrically neutral. It will be seen later that care has to be taken in applying this condition of electrical neutrality at or near a junction between different materials. In general, it will be found that different materials have to have a *built-in* electric potential between them so as to establish the equilibrium concentrations of charge carriers away from the junctions. This built-in potential requires a deviation of electrical neutrality at or near the junction. There is another area where electrical neutrality is not preserved—in non-linear conductors. In particular, if the mobility is a function of the electric field such that dv/dE is negative over some range of fields then charge will spontaneously accumulate rather than disperse. This will be met in discussions on the Gunn effect (Chapter 8).

Displacement current

Gauss' theorem leads one to consider a form of electrical conduction that does not rely on the direct movement of charge within a solid. Consider, for example, a parallel-plate capacitor of area A and plate separation d with a current I flowing into one of the plates; the charge q on this plate must be accumulating at a rate $dq/dt = I$. Equally, on the other plate there must be a negative charge $-q$ which is decreasing at the same rate. The current *in* thus equals the current *out*. Gauss' theorem shows that the electric field between the plates is $E = q/A\varepsilon_0\varepsilon_r$ so that $I = A\varepsilon_r\varepsilon_0 \, \partial E/\partial t$. The time-varying field thus induces charge on the opposite plate so as to keep the current flow continuous and may itself be looked upon as contributing a current density across the plate. This form of current is called the *displacement* current because the electric flux vector $\mathbf{D} = \varepsilon_r\varepsilon_0\mathbf{E}$ is also called the displacement vector. Thus the movement of charge creates a convection current density $\mathbf{J}_c$ while the time-varying fields create a displacement current density.

The two components combine to give a total current density $\mathbf{J}_t$

$$\mathbf{J}_t = \mathbf{J}_c + \varepsilon_r \varepsilon_0 \, \partial \mathbf{E}/\partial t \qquad\qquad 3.11$$

at any point. In three dimensions, Maxwell's equations for electromagnetic theory state that curl $\mathbf{H} = \mathbf{J}_t$ while in one dimension it may be shown that the total current is independent of position (Problem 3.3). *The total current is always the terminal current in any 'one-dimensional' device.*

Strictly speaking, there is always a displacement current component in any resistor, i.e. any real resistance must have some associated capacitance. However, in Problem 3.2 it is shown that when an electric field is changing at a frequency ω, the peak displacement current is $\omega \tau_d$ times the peak conduction current, where τ_d is the dielectric relaxation time. Thus, provided that τ_d is small as already indicated, one may neglect the displacement current for many problems. It is of course zero in the steady state where there is no dependence of the fields upon time.

3.4 Diffusion

The motions of charges inside a conductor or a semiconductor have so far been likened to those of gas molecules where the motion is usually quite random; only if there is a net movement of charge in some particular direction is current said to flow. Although the electric field is one obvious agent for causing a net movement of the charge carriers, we consider here a spatial gradient in the charge carrier density. In Fig. 3.2, there is sketched a junction between materials with charge-carrier densities n_1 and n_2/m^3 respectively. Essentially half of the charge carriers move one way and the other half the other way, the velocity of movement being directly related to the thermal energy. There is then a current density that is proportional to $D_1 n_1$ flowing from material 1 to 2 and $D_2 n_2$ flowing from 2 into 1 where D is some measure of the random thermal velocity. For small changes of n_2 from n_1 one finds that the net particle flux density is given by Fick's law as $-(\partial/\partial x)(Dn)$ where

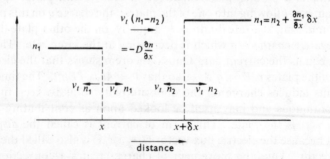

Fig. 3.2 Mechanism of diffusion. v_t is a thermal velocity $v_t \sim (kT\mu/e)$.

D is the diffusion constant and the minus sign indicates that the movement is *from* the region of higher particle density *to* the lower. Usually the temperature of the material is uniform, so that the diffusion constant is also uniform across the gradient of charge carriers; the diffusion current density for electrons is then given by

$$J_{dn} = (-e)\left(-D_n \frac{\partial n}{\partial x}\right) = eD_n \frac{\partial n}{\partial x} \qquad 3.12a$$

while for holes it is given by

$$J_{dp} = e\left(-D_p \frac{\partial p}{\partial x}\right) = -eD_p \frac{\partial p}{\partial x} \qquad 3.12b$$

The diffusion constants D_n and D_p must not be confused with the impurity atoms' diffusion constants which are completely negligible in value at normal operating temperatures for most impurities. It will be shown later that, for charge carriers in many semiconducting materials, $D = \mu kT/e$ so, for mobilities around 0·1 to 1 m^2/V.s, one finds values for the diffusion constant at 300 K around $25 . 10^{-4}$ to $25 . 10^{-3}$ m^2/s.

This diffusion current must be added to any 'Ohmic' current to give the complete *convection current* density as

$$J_{cn} = en\mu_n E + eD_n \frac{\partial n}{\partial x} \qquad \text{(for electrons)} \qquad 3.13a$$

or

$$J_{cp} = ep\mu_p E - eD_p \frac{\partial p}{\partial x} \qquad \text{(for holes)} \qquad 3.13b$$

The Debye length

We can now begin to see what will happen at interfaces between two pieces of the same material with different impurity concentrations. Consider for the moment (Fig. 3.3) relatively small changes of the ionized acceptor concentrations say from P_2 to P_1. From electrical neutrality one would expect there to be P_2 and P_1 holes respectively in the materials (we are assuming extrinsic semiconductors with negligible numbers of carriers of the opposite polarity). If, however, this solution were maintained up to the junction then there would be a sharp concentration gradient of the charge carriers which would be free to diffuse from the higher to the lower concentration. At first sight it might be thought that this diffusion would continue to drive mobile charges from the higher concentration to the lower—like a stream of water seeking the lowest level. However, this would imply a current flow even in the absence of an applied potential—a result that is refuted by experiment. In equilibrium with no applied potentials the current

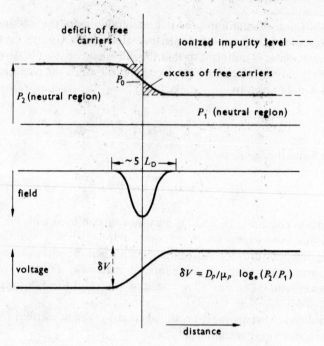

Fig. 3.3   Readjustment of charge at an interface.

flow is zero: $J_{cp} = 0$. However in the region labelled (1) the net charge density may be put as $\rho = e(p - P_1)$ with the fixed ionized acceptor charge as $-\rho_0 = -eP_1$. The fixed charge is uniform with distance over the region 1. From eqns 3.7 and 3.13b one may eliminate the field E, so that one obtains, using $J_{cp} = 0$,

$$\rho = L_d{}^2 \frac{\partial}{\partial x}\left(\frac{1}{\rho_0 + \rho} \frac{\partial \rho}{\partial x}\right) \qquad\qquad 3.14$$

where $L_d{}^2 = (D_p \varepsilon_r \varepsilon_0 / \rho_0 \mu_p)$. The characteristic length L_d is called the *Debye length* and appears in other work elsewhere. Provided that at the boundary $\rho = e(P_0 - P_1)$ with ρ small compared to ρ_0 then a solution is given by

$$\rho \simeq e(P_0 - P_1) \exp(-x/L_d) \qquad\qquad 3.15$$

A similar result can be derived for the region (2) with the charge carriers decaying away from the value P_2 to P_0 at the boundary with an appropriate characteristic distance corresponding to the Debye length for that material.

Note that for a few Debye lengths around an abrupt change in the donor or acceptor concentrations electrical neutrality does not hold. It will be seen later (problem 3.5) that overall electrical neutrality holds, any excess on one side equals any deficit on the other, but *local* neutrality does not hold. However one will find that typical distances for the Debye length, in materials of interest, are around 10^{-7} m or less and the shortness of this scale of distance can often be used as the basis of simplifying practical problems.

The built-in potential

The condition that $J_{cp} = 0$, in eqn 3.13b, can be used in another way combined with $E = -\partial V/\partial x$. One can then show that

$$\frac{\partial V}{\partial x} = -(D_p/\mu_p)\frac{1}{p}\frac{\partial p}{\partial x} \qquad 3.16$$

Integrating from $p = P_2$ to P_1, one finds that the fields arising from the local lack of charge neutrality give rise to a potential

$$\delta V = (D_p/\mu_p)\log_e(P_2/P_1) \qquad 3.17$$

where the sense of this potential is such as to repel the charge carriers from diffusing any more from the higher to lower concentrations (i.e. in this case with $P_2 > P_1$ the 1 region is positive with respect to the 2 region). This built-in potential is also referred to as the *diffusion potential*.

It can be seen from the model given here that this potential has been formed from a redistribution of the charge, this redistribution being demanded by the condition that *no* current flows. One cannot therefore utilize this potential to drive any current. Any attempt to do so, such as joining the materials with a wire, will only lead to further redistributions of charge that will prevent current flowing unless an external potential is applied. Later sections shows how this diffusion potential can be described more elegantly through the concept of the *Fermi level*.

3.5 The Fermi level: the Fermi-Dirac function

Statistical thermodynamics shows us that, when we are considering large numbers of particles or systems which interact, we can calculate the probability of one particle or system being at an energy $\mathscr{E}$ from general reasoning about the numbers of ways in which the energy can be distributed among the particles. There are many treatments at an elementary level (see references 3.3 to 3.7) and so the study of this area is left for parallel reading. In a classical discussion, where the particles are assumed to behave like the components of an ideal gas, the probability of a state at an energy $\mathscr{E}$ being occupied is given by the Boltzmann distribution (Fig. 3.4a):

$$F(\mathscr{E}) = \exp(-\mathscr{E}/kT).$$

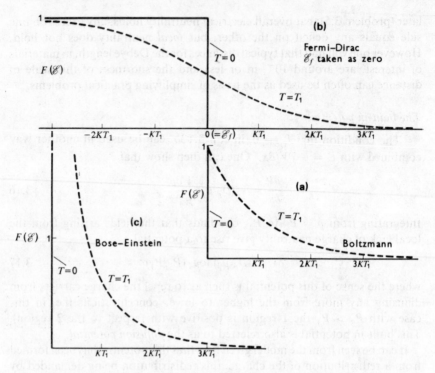

Fig 3.4 Energy distributions. $F(\mathscr{E})$ is the probable occupation number of quantum states.

The higher the energy the less likely is the state to be occupied. The Boltzmann distribution implies that at zero temperature the only energy state that can be occupied is the zero energy state, $\mathscr{E} = 0$, though of course there may be some reference energy added to this zero without radically altering the basic probability distribution. However, if this classical distribution is applied to the electrons in a metal, one immediately sees that at 0 K the Pauli exclusion principle is violated. The lowest possible energy state for the electrons in a metal is the one where all the quantum states in the metal are filled with just one electron ranging from the lowest energy state up to some energy $\mathscr{E}_f$ when all the electrons have been accommodated. The level $\mathscr{E}_f$ is thus determined by the numbers of electrons to be accommodated, along with the quantum state structure of the metal. One can say that for all energies below $\mathscr{E}_f$ the probability of occupation at $T = 0$ K is unity, while for all those states above $\mathscr{E}_f$ the probability of occupation is zero. One then finds that the Pauli exclusion principle limits the number of ways the electrons can be distributed and this leads to a modification of the probable energy distribution. The new energy distribution is called the Fermi-Dirac distribution (Fig. 3.4b) and is given by:

$$F(\mathscr{E}) = 1/[\exp{(\mathscr{E} - \mathscr{E}_f)}/kT + 1] \qquad 3.18a$$

The reader should check that this gives the desired form at $T = 0$ K. The energy $\mathscr{E}_f$ in this function is then called the *Fermi level;* it can be seen that when $\mathscr{E} - \mathscr{E}_f \gg kT$ then

$$F(\mathscr{E}) = \exp{-(\mathscr{E} - \mathscr{E}_f)}/kT \qquad 3.18b$$

so that the Fermi level acts as a reference energy for this approximation, which is referred to as the *Boltzmann approximation.*

Quantum statistics are slightly easier to follow than classical statistics because of the well-developed concept of a quantum state. It is then easier to consider permutations of particles in such well-defined states. When the Pauli principle does not hold, as for example with photons which are the particles associated with electromagnetic radiation, one finds that the possible combinations lead to an energy distribution $F_{be}(\mathscr{E}) = [\exp{(\mathscr{E}/kT)} - 1]^{-1}$. This is the Bose Einstein distribution [Fig. 3.4c] and again leads to the Boltzmann distribution at high enough energies. However at really low energies the probable number of particles in an energy level at $\mathscr{E}$ exceeds the number of states since $F_{be}(\mathscr{E}) > 1$ if $\mathscr{E} < (\ln 2) kT$.

In order to give the reader some immediate idea of the use of the Fermi level and this statistical approach to obtaining equilibrium conditions, we shall now consider two metals A and B that are joined together but otherwise isolated (Fig. 3.5). The Fermi levels in each of these metals, when considered separately, are at E_{fa} and E_{fb} at an energy ϕ_a and ϕ_b below the *vacuum level.* This latter level indicates the energy required for the electron to become free of the metal. Thus ϕ_a is the minimum energy required by an electron to escape from the metal A and be *emitted* from the surface. Indeed *thermionic emission,* the phenomenon of emission of electrons form very hot metals

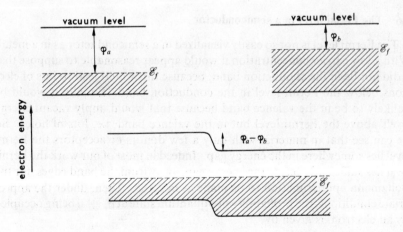

Fig. 3.5   Fermi levels in metals.

such as heated filaments of tungsten, is caused by the tail of energies on the Fermi-Dirac distribution allowing some energies to exceed ϕ, so that these electrons can escape. The energy ϕ is known as the *work function*. When the isolated metals A and B come into contact one can say either that the electrons from the metal with the lower work function are emitted into the metal with the higher work function or, equally, that the electrons diffuse from the higher energies into the lower energies. The process of diffusion or emission must continue until the potential of one metal rises enough to stop the process of diffusion. This will happen when the probability of occupation of the energy levels on either side of the junction is the same; this means that the Fermi levels must be uniform throughout the combined material. There will then be a built-in potential difference between the two metals given by $\phi_b - \phi_a$, the difference of work functions. The metals, being good conductors, remain substantially neutral, with a charge distribution very close to the interface creating the required potential difference.

As for the different semiconductors, the built-in potential cannot be made to do work. If another metal C is connected to A and B then the Fermi level will align in C with that of A and B and, after an initial redistribution of charge, no current will flow, unless some potential is applied. When a potential V is applied between materials it is the Fermi levels which have to change their separation by V—thus again the Fermi level is the reference. Heat can create an e.m.f. which will drive a current around a metal circuit, as in a thermo-couple. In this case the electrons diffuse preferentially from one metal to the other with the heat supplying the work to drive any current. However, in our work, the junctions can be considered to be at a uniform temperature.

3.6   The Fermi level in a semiconductor

The Fermi level is not so easily visualized in a semiconductor as in a metal. With a low carrier concentration it would appear reasonable to suppose that it did not lie in the conduction band, because this would lead to lots of electrons below the Fermi level in the conduction band. Equally it would be unlikely to be in the valence band because that would imply vacant energy levels above the Fermi level but in the valence band, i.e. lots of holes. So we can see that in material with only a few donors or acceptors the Fermi level lies somewhere in the energy gap. Indeed in most of our work the Fermi level lies sufficiently inside the energy gap, away from the band edges, for the Boltzmann approximation to be made as in 3.18b. Then, under the appropriate conditions, the probability of a quantum state at $\mathscr{E}_c + u$ being occupied by an electron, is given by

$$F(\mathscr{E}_c + u) \simeq [\exp -(\mathscr{E}_c - \mathscr{E}_f)/kT] \exp (-u/kT) \qquad 3.19$$

while similarly for the probability of a hole occupying a quantum state at $\mathscr{E}_v - u$ is given by

$$[1 - F(\mathscr{E}_v - u)] \simeq [\exp -(\mathscr{E}_f - \mathscr{E}_v)/kT] \exp(-u/kT) \qquad 3.20$$

We now have to turn to the concept of density of states. In the conduction band, between the energies $\mathscr{E}_c + u$ and $\mathscr{E}_c + u + \delta u$ there will be many quantum states. The number of states will be proportional to δu, say $N(u)\,\delta u$ per unit volume of material. The explicit form for this *density of states* $N(u)$ is not really needed for the present argument and will be left until Chapter 6. The density, n, of the conduction electrons in the material is given by the sum of all the quantum states in the conduction band weighted by their probability of occupation with an electron, given from eqn 3.19. The summation is in fact done by an integration:

$$n = [\exp -(\mathscr{E}_c - \mathscr{E}_f)/kT] \int_{u=0} N(u) \exp -u/kT\,\mathrm{d}u$$
$$= N_c \exp -(\mathscr{E}_c - \mathscr{E}_f)/kT \qquad 3.21$$

where the integral has been evaluated as N_c. We shall be able to show later that N_c has a numerical value around $10^{25}/m^3$ for many semiconductors. The density of all the states, N_c, is called the *effective density of states*. It indicates that the electrons can be treated on a statistical basis as if there were N_c quantum states per unit volume of material and all these states lay at the energy $\mathscr{E}_c$. An exactly analogous piece of work can be carried out replacing $\mathscr{E}_c$ by $\mathscr{E}_v$, $+u$ by $-u$, N_c by N_v and the term *electrons* by *holes*. The density of holes in the valence band is then given by

$$p = N_v \exp -(\mathscr{E}_f - \mathscr{E}_v)/kT \qquad 3.22$$

and N_v is the effective density of states for the valence band. It has typically an order of magnitude similar to that of the effective density of states for electrons. As for electrons the result allows one to consider all the holes effectively lying at the energy $\mathscr{E}_v$ in N_v quantum states per unit volume.

The result immediately shows that Boltzmann statistics can only be applied if the actual number of electrons in the conduction band is less than N_c or if the number of holes in the valence band is less than N_v (per unit volume). Any attempt to insert more carriers than this must push the Fermi level into the conduction or valence bands. Material with such a high concentration of carriers requires the full Fermi-Dirac probability function to describe the statistics of occupation of the quantum states with accuracy. The material is then referred to as *degenerate;* material to which the Boltzmann approximation can be applied is termed *non-degenerate*. In most of this book, the material is taken to be non-degenerate for any calculations.

The position of the Fermi level for intrinsic semiconductor material can now be evaluated. In the intrinsic material there are no impurities, so that for every electron which breaks its valence bond, and moves into the conduction band, there is a hole left in the valence band. Charge neutrality is assured, so $n = p$. Equating the results of eqns 3.21 and 3.22 yields, after a little rearrangement,

$$\mathscr{E}_f = \tfrac{1}{2}(\mathscr{E}_c + \mathscr{E}_v) + \tfrac{1}{2} kT \log_e (N_v/N_c) \qquad 3.23$$

Thus for modest temperatures the Fermi level lies close to the middle of the band gap, or even at the middle, if $N_v = N_c$.

The results of eqns 3.21 and 3.22 are valid for both extrinsic and intrinsic materials, so that by multiplication one obtains

$$pn = n_i^2 \quad \text{with} \quad n_i^2 = N_c N_v \exp -(\mathscr{E}_c - \mathscr{E}_v)/kT \qquad 3.24$$

The quantity n_i is the *intrinsic concentration*, being the density of holes and of electrons for intrinsic material. The value of n_i is exceedingly small for good *extrinsic* semiconductors so that in equilibrium with a reasonable number of charge carriers of one type (*majority carriers*) neglecting the charge carriers of the opposite type (the *minority carriers*) is often justified.

In extrinsic semiconductors the charge carriers are inserted by the addition of impurities. Suppose that there are N_d donor impurities per unit volume, giving rise to energy levels at $\delta\mathscr{E}$ below the conduction-band energy $\mathscr{E}_c$. Assuming that the Boltzmann approximation holds, one has

$$N_d \exp -(\mathscr{E}_c - \delta\mathscr{E} - \mathscr{E}_f)/kT$$

electrons occupying the N_d donor states.† In other words the donors have donated $N_d[1 - \exp(\mathscr{E}_f + \delta\mathscr{E} - \mathscr{E}_c)/kT]$ electrons to the conduction band along with p electrons that have been thermally stimulated from the valence band, leaving behind p holes in the valence band. Electrical neutrality and equilibrium will then imply that

$$p + N_d[1 - \exp(\mathscr{E}_f + \delta\mathscr{E} - \mathscr{E}_c)/kT] = n$$

where 3.24 and 3.22 can be used to eliminate p and $\mathscr{E}_c - \mathscr{E}_f$

$$n_i^2 + nN_d[1 - (N_d/N_c) \exp \delta\mathscr{E}/kT] = n^2 \qquad 3.25$$

Provided that $n_i^2 \ll N_d^2$ then

$$n \simeq N_d/[1 + (N_d/N_c) \exp \delta\mathscr{E}/kT] \qquad 3.26a$$

---

† There are two possibilities for donor or acceptor states: they can have one electron but it can be of either spin. This is shown in A. H. Wilson, *Theory of Metals* (Cambridge University Press), pp. 326–8, to give a modified Fermi distribution $1/[1 + \tfrac{1}{2} \exp(\mathscr{E}_f - \mathscr{E})/kT]$. Thus $N_d[1 - 2 \exp -(\mathscr{E}_c - \delta\mathscr{E} - \mathscr{E}_f)/kT]$ is the approximate number of ionized donor states. In our work the factor makes little difference because $n \sim N_d$ is always assumed to be true.

From eqn 3.21

$$\mathscr{E}_f = \mathscr{E}_c - kT \log_e (N_c/n) \qquad 3.26b$$

This then is a basic calculation for estimating the Fermi level in a non-degenerate extrinsic semiconductor with a known donor concentration N_d. Observe from eqn 3.26a that, provided $N_d \exp \delta\mathscr{E}/kT \ll N_c$, then $n \sim N_d$, so that virtually all the donors are ionized. This is the required result for non-degenerate material. As N_d approaches the value of N_c so the full Fermi-Dirac distribution must be used.

The corresponding results for holes should now be derived by the reader. These give

$$p = N_a/[1 + (N_a/N_c) \exp \delta\mathscr{E}/kT] \qquad 3.27a$$

$$\mathscr{E}_f = \mathscr{E}_v + kT \log_e (N_v/p) \qquad 3.27b$$

where the acceptors, N_a per unit volume, give rise to a level at $\mathscr{E}_v + \delta\mathscr{E}$.

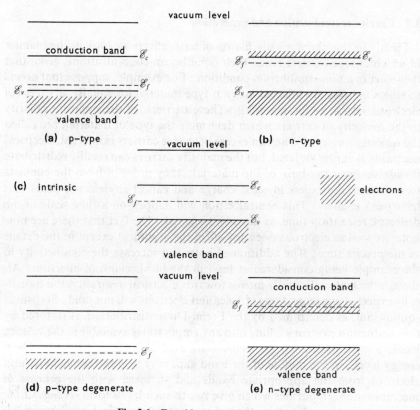

Fig. 3.6 Fermi levels in semiconductors.

Figure 3.6 indicates the scheme of the Fermi levels for the different types of semiconductor that will be considered. Note that the material designated p^+ and n^+ is usually degenerate or nearly degenerate, so that Boltzmann statistics give a poor approximation to reality. The Fermi level lies close to, or indeed within, the allowed bands for the electrons. However, such material is often merely contact material to the more lightly doped types, and the qualitative description given by the Boltzmann approximation for this degenerate material can still be sufficiently accurate for our purposes.

In summary, the Fermi level is the reference level. Any changes of external potential appear as changes in the Fermi level. In equilibrium with no *applied* potentials the Fermi level is constant everywhere and equilibrium is only established when the charges have arranged themselves in a manner that is consistent with this alignment of Fermi levels. This concept is of great value in calculating equilibrium conditions, but the calculations for semiconductor junctions are left for Chapter 4.

3.7 Carrier recombination and generation

Finally in this chapter's discussion of basic effects we look at the manner in which charge carriers reach their equilibrium concentrations, given that they start in a non-equilibrium condition. For example, suppose that excess numbers of holes are injected into a n-type material (or equally an excess of electrons into a p-type material). These carriers, of the opposite polarity to the majority of carriers which determine the type of material, are called the *minority carriers*. When an excess of charge carriers is injected, electrical neutrality is at first violated, but the majority carriers can readily redistribute themselves over the bulk of the material; they move in from the contacts to neutralize the excess injected charge and cancel any electric fields that have been created. This neutralization will happen on a time scale of the dielectric relaxation time, as already discussed. The fact that there are now holes as well as electrons does not affect the argument except in the details of the precise time. The addition of holes can increase the conductivity in the example being considered at least in the local region of injection. Although the material rapidly moves towards electrical neutrality, the statistically expected concentrations of holes and electrons will not hold. Statistical equilibrium, as determined by the Fermi-Dirac distribution, is restored by the conduction electrons falling into any empty states available in the valence band; a process known as *recombination* of holes and electrons. Deep energy levels, in the middle of the band gap, may help to receive holes and electrons from the appropriate bands and so assist with the process of recombination. Impurities which give rise to such levels form *recombination centres*. It is recombination then that restores statistical equilibrium by removing excess holes and electrons. Just as recombination removes excess carriers so the process of thermal agitation breaking the valence bonds and *generating* charge carriers can restore the balance when there is a deficit of carriers. It must be emphasized that the conduction electrons and valence-band holes are continuously recombining and being generated. Equilibrium is a dynamic state where the generation rate G just balances the recombination rate R. The time scale required for the readjustment of the carrier densities by these processes is usually governed by a *minority carrier lifetime*, denoted here by the symbol τ_r. Its value typically varies from several microseconds down to nanoseconds, depending on materials and the impurities.

One method of measuring this time can be made using the following ideas. It is known that light of frequency f consists of quanta of energy hf. If this energy is sufficient to break a valence bond then it may create a hole and electron pair that contribute to the conduction. Thus subjecting a bar of semiconductor to a modulated light source can modulate the conductivity by hole–electron generation. When the light is turned off the conductivity can be noted to decay with a time constant determined by τ_r, the recombination time. In all the work in this book it will be assumed that τ_r is very much longer than the dielectric relaxation time of the semiconductor. In this case electrical neutralization occurs more rapidly than statistical equilibrium. In an appendix we briefly consider some consequences of having material where the converse is true. This type of material has been referred to as 'relaxation' semiconductor material because its physics is dominated by the long dielectric relaxation time.

In Appendix on recombination, a standard account of the more detailed aspects of the effect is given, and the effects of deep impurity levels are considered. In this chapter we present a much simpler though less general example, where direct recombination from the conduction band to the valence band is assumed. Generation is the exact converse. As the estate agent knows well, the number of house sales is proportional to both the number of buyers and to the number of sellers. This is an example of the law of mass action which states that the rate of reaction is proportional to the concentration of the components taking part. The recombination rate R is thus proportional to both the number of electrons in the conduction band and to the number of holes in the valence band, so that one may write

$$R = R_e(np/n_i^2) \qquad 3.31$$

The equation has been normalized so that R_e is the equilibrium rate when $np = n_i^2$. The process of generation, is by the same arguments, proportional to the number of electrons *in the valence band* available for ionization, and to the number of vacant sites *in the conduction band* available to receive them.

These numbers are, however, only marginally altered by changes in the numbers of free charge carriers, at least for non-degenerate semiconductors. Thus the generation rate G can be written as G_e, a constant. In equilibrium, G_e and R_e must cancel each other, so that the total net rate of recombination for the carriers is given by:

$$R - G = R_e[(np/n_i^2) - 1] \qquad 3.32$$

Now put $n = n_e + \delta n$ and $p = p_e + \delta p$ and suppose that because of the fast dielectric relaxation time electrical neutrality keeps $\delta n = \delta p$; then, to a first order in $\delta p (= p - p_e)$, one has

$$R - G = (p - p_e)/\tau_{rp} \quad \text{with} \quad \tau_{rp} = n_i^2/R_e(n_e + p_e) \qquad 3.33$$

The suffix e denotes equilibrium values. The parameter τ_{rp} is termed the minority carrier lifetime for holes. In the more general treatment given in the appendix the minority carrier lifetime for holes and electrons can be distinctly different theoretically, more in accord with experimental findings. In this simple treatment, τ_{rn} and τ_{rp} are found to be the same.

The charge continuity eqn 3.8 now applies to both holes and to electrons provided that allowance is made for recombination and generation. For example if, in Fig. 3.1b, the charge ρ is given by the hole concentration, ep, then one must allow for the loss of holes caused by recombination at a rate $-[e(p - p_e)/\tau_{rp}]A \, \delta x$ from the element of volume $A \, \delta n$, thus for the holes separately one must have

$$\frac{\partial ep}{\partial t} = -\frac{\partial J_{cp}}{\partial x} - \frac{e(p - p_e)}{\tau_{rp}} \qquad 3.34a$$

while for the electrons separately one must have, by similar arguments,

$$-\frac{\partial en}{\partial t} = -\frac{\partial J_{cn}}{\partial x} + \frac{e(n - n_e)}{\tau_{rn}} \qquad 3.34b$$

The form of eqn 3.34 still applies whether the carriers are majority carriers or minority carriers, but when both are considered together (i.e. $\rho = ep - en$ and $J = J_{cp} + J_{cn}$ in Fig. 3.1b) then one must be able to add both parts of eqn 3.34 to form eqn 3.8. The recombination rates for the holes and electrons are then equal and give opposite charge at any instant—the *minority* charge carriers usually determining the recombination rate. This then concludes the outline of basic effects.

PROBLEMS

3.1 By consideration of a cube of sides $\delta x, \delta y, \delta z$, extend the theory given in section 3.3 to show that

$$\frac{\partial E_x}{\partial x} + \frac{\partial E_y}{\partial y} + \frac{\partial E_z}{\partial z} = \rho/\varepsilon_r\varepsilon_0$$

and similarly

$$\frac{\partial J_x}{\partial x} + \frac{\partial J_y}{\partial y} + \frac{\partial J_z}{\partial z} + \frac{\partial \rho}{\partial t} = 0$$

Hence demonstrate that eqn 3.10 is not dependent upon a one-dimensional current flow.

3.2 A resistor is constructed of a material with a resistivity ρ and has an area A and length L. The relative dielectric constant is ε_r for the material. If C is the capacitance between the terminals and R is the resistance, show that $RC = \tau_d$ where τ_d is the dielectric relaxation time $\varepsilon_r\varepsilon_0\rho$. Hence show that the capacitive current is ω/ω_c times the conduction current in such a resistor. $[\omega_c\tau_d = 1]$.

3.3 From eqns 3.7 and 3.8 and using $\partial^2/\partial x\,\partial t = \partial^2/\partial t\,\partial x$ show that $\partial/\partial x(J_t) = 0$ where J_t is defined by eqn 3.11.

Hence show that J_t. Area is the current through the terminals to any device. [*Aliter*: note that curl $\mathbf{H} = \mathbf{J} + \partial\mathbf{D}/\partial t$ and Div (curl $\mathbf{H}$) $\equiv 0$]

3.4 Discuss the variation of the Fermi level with temperature. Show that as the temperature rises the Fermi level moves towards the intrinsic position for the Fermi level in the material, and as the temperature falls the Fermi level moves towards the impurity level which determines the sign of the charge carriers. In Si (effective density of states approximately $2 \cdot 8 \times 10^{25}/m^3$) there are $10^{21}/m^3$ arsenic impurity atoms (ionization energy approximately 50 mV). Show that at 75 K the free-electron concentration will have fallen approximately to one half that at room temperature. Where will the Fermi level be at this temperature?

3.5 Using Gauss, show that the equality of the fields deep in regions 1 and 2 of Fig. 3.3 implies that the charge excess is equal to the charge deficit around the junction. Hence show that with the approximations of eqn 3.15

$$(P_0 - P_1)L_{d1} = (P_2 - P_1)L_{d2}$$

where L_{d1} and L_{d2} are the Debye lengths in the regions 1 and 2 respectively. Hence determine P_0 and show that the peak electric field is $e(P_2 - P_1)L_{d1}L_{d2}/\varepsilon_0\varepsilon_r(L_{d1} + L_{d2})$.

General references

KITTEL, C. Reference 2.1.
3.1 SMITH, R. A. *Semiconductors*. Cambridge University Press, 1959.
3.2 BLATT, F. J. *Physics of Electronic Conduction in Solids*. McGraw-Hill, 1968.

Special references

Statistical physics and conduction theory

3.3 ANDREWS, F. C. *Equilibrium Statistical Mechanics*. Wiley, 1963.
3.4 POINTON, A. J. *Introduction to Statistical Physics*. Longmans, 1967.
3.5 REIF, F. *Statistical Physics*. Berkeley Physics Course, Vol. 5. McGraw-Hill, 1967.
Good introduction to the general ideas and concepts of statistical physics but omits quantum statistics.

3.6   BORN, M. Reference 2.8: Chapter 7, Quantum Statistics.
3.7   HUANG, K. *Statistical Mechanics*. Wiley, 1963.
          In general, very advanced, but gives a helpful unified derivation of Boltzmann, Bose-
          Einstein and Fermi-Dirac distributions on p. 194.
3.8   CARROLL, J. E. Reference 8.1. Chapter 2 gives an elementary discussion of relaxation time
          approach to solution of Boltzmann collision equation.

Electromagnetic theory

3.9   RAMO, S., WHINNERY, J. R. and VAN DUZER, T. *Fields and Waves in Communication Electronics*.
          Wiley, 1965.

4

p-n Junctions

4.1 A practical p-n junction

This chapter develops the theory of the p-n junction and gives a more detailed account of its uses than the outline given in Chapter 1. To make the discussion less abstract, one particular construction, which has practical importance, will be outlined. The process is roughly as follows: a slice is cut from a boule of a highly doped semiconductor, say n-type silicon. This will give a disc of material which is several centimetres in diameter and about a millimetre thick. One face of this slice is polished to a mirror finish before placing in an epitaxial reactor where an epilayer of n-type material is grown to a depth of a few micrometres. This epilayer is usually of much greater purity than the substrate starting slice. After growth, the epitaxial layer is assessed for its mobility, donor density and other qualities, though usually only a special control slice is assessed (a few techniques of assessment are outlined in Appendix 2). Into the surface of this epilayer is then diffused a p-type impurity which overdopes the n-type impurity already present. A typical impurity profile for this result is shown in Fig. 4.1a. This is the p^+-n-n^+ structure where the superscripts $+$ denote a very heavy doping, usually around 10^{24} impurities/m^3. The slice will be thinned down by lapping the unwanted side with a fine alumina powder or the equivalent. The large circular discs with this structure can then be cut-up or photo-engraved into thousands of separate chips, each forming a diode. There will be additional steps, not mentioned in this outline, such as photolithography and diffusion masking to define the diffusion areas and the contacts to the diodes. Finally there is the metallization process to make the electrical contacts. These contacts can be fairly readily made to the highly doped p^+ and n^+ materials by alloying suitable metals into their surfaces.

The important features of this structure will be seen, during the course of this and other chapters, to be its versatility. Merely by changing the length and doping of the central n-type region one will be able to change from a high-voltage rectifier through to an ultra-high-frequency device with a negative resistance. The highly doped sections are really providing low electrical resistance contacts to the middle section which does most of the

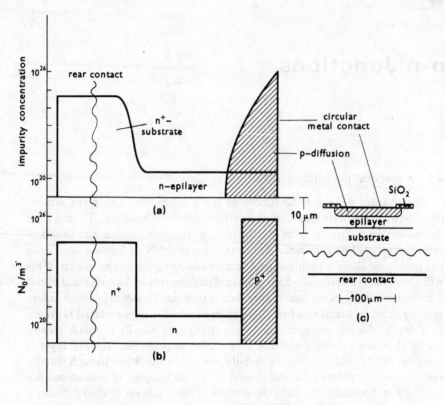

Fig. 4.1 The p$^+$-n-n$^+$ diode. (a) Typical profile: epilayer $\sim$1–100 μm thick, substrate $\sim$100 μm thick, contact diameters $\sim$10–10,000 μm depending on application. (b) Idealized abrupt junction considered analytically. (c) Fabrication with diffusion through oxide mask (vertical scale expanded).

work and determines the properties of the diode. The substrate also provides a more solid backing for the very thin active region and so makes the device easier to handle. Other structures can of course be made, in particular the complementary n$^+$-p-p$^+$ structure. In some diodes where only high-impurity concentrations are required there will be no need for any middle n or p section.

The chapter starts with a more detailed account of the physics of equilibrium and current flow between a p and n material. Discussions on specific devices are then presented. The last two sections (4.6 and 4.7) cover rather specialized devices and may be omitted on a first reading without much detriment to the following chapters. It is, however, probably the specialized applications discussed in sections 4.5 to 4.7 that will ensure a future for the p-n junction diode; its use as a mere rectifier or even variable capacitance must be compared with similar uses of a metal contact to a semiconductor (Schottky barrier) discussed in Chapter 7.

4.2   The alignment of the Fermi level in p-n junctions

Like many practical problems, progress in understanding is made most rapidly by judicious simplification. The structure of the doping profile shown in Fig. 4.1a is idealized into the abrupt changes shown in Fig. 4.1b. The Boltzmann approximation (3.18b) will be assumed to hold, though, at the doping levels indicated by a normal p^+ or n^+ impurity, the full Fermi-Dirac statistics should be used for maximum rigour. However the qualitative ideas would not be altered in any way.

Figure 4.2a indicates a layer of lightly doped n-type material sandwiched between two more heavily doped sections. The appropriate Fermi level in each section is shown in Fig. 4.2b, with transition regions where the charge is redistributed so as to bring about a change of potential to align the Fermi level throughout the material. The n^+-n junction need not be considered in detail because this is merely a contact to the important n-type region from the substrate; its behaviour follows qualitatively that described for the junction between two materials with different p-type doping (section 3.4). The transition region between the p and n materials is more interesting. The electrons diffuse out of the n-type region towards their low concentration in the p-material, and similarly the holes diffuse out of the p-material towards

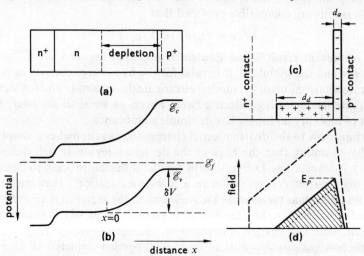

Fig. 4.2 The p-n junction. (a) Structure. (b) Fermi level and band structure in equilibrium. (c) Idealized charge distribution of un-neutralized impurities. (d) Field profile in equilibrium and above punchthrough ___ and also on forward bias....

the low concentrations in the n-material. This diffusion lowers rapidly the concentration of the mobile charge carriers in the respective regions, to leave a zone that is relatively depleted of mobile carriers but has a background of ionized impurities. Thus this depletion region is electrically charged; the charge adjusts, by a change in the depletion width, to the amount that is required to bring about the alignment of the Fermi level throughout the semiconductor, as determined by the statistical approach.

For the moment the precise distribution of charge need not be considered as we need only determine the potential δV (as in Fig. 4.2b) that is required to align the Fermi levels. The Fermi levels can be found with respect to the band edges from eqns 3.26b and 3.27b for the p- and n-materials separately. The potential δV is then the difference between the two values:

$$e\,\delta V = E_c - E_v - kT \log_e (N_c/N_d) - kT \log_e (N_v/N_a)$$

but using eqn 3.24 to eliminate $E_c - E_v$ one obtains

$$\delta V = (kT/e) \log_e (N_d N_a/n_i^2) \qquad\qquad 4.1a$$

A similar result can be found by considering the diffusion of the N_a holes from the p-material over into the n-material which has only n_i^2/N_d holes. Using eqn 3.17 with the appropriate numbers, one can find that

$$\delta V = (D_p/\mu_p) \log_e (N_d N_a/n_i^2) \qquad\qquad 4.1b$$

An exactly analogous process occurs for the electrons diffusing from the n-type into the p-type material but now (D_n/μ_n) replaces (D_p/μ_p) in 4.1b. All three results are compatible provided that

$$kT/e = D_p/\mu_p = D_n/\mu_n \qquad\qquad 4.2$$

This important result is the *Einstein relationship* between the diffusion coefficients and the mobility. It is valid for the free charge carriers in non-degenerate semiconductors at modest electric fields. It is not valid for highly doped materials or at high electric fields where, as we shall see later, the concept of mobility no longer has its simple significance.

Returning now to the distribution of charges, one again makes a simplification by assuming that the edge of the depletion region is well defined. In reality it takes a few Debye lengths for the diffusion process to change from a high carrier concentration to a low concentration. However it is frequently found that the average Debye length in the material is an order of magnitude smaller than the transition or depletion region, so it is reasonable to assume that the carrier concentration changes abruptly (Fig. 4.2c). The depletion regions are not, of course, altogether depleted of charge carriers; the concentration of the free carriers is merely much less than that of the fixed ionized centres. Thus the charge is virtually that of the ionized impurities. The physical picture is now set for some simple calculations.

The essential process of adjustment of the depletion region is caused by any field at the edges of the depletion region moving the charge carriers until the required potential δV is achieved. One can say that once the equilibrium situation has been established the electric field is zero at the edges of the depletion zone. Gauss' theorem, applied to the n-side, yields

$$\varepsilon_0 \varepsilon_r \frac{\partial E}{\partial x} = eN_d$$

where N_d is the ionized donor density (positive charge), with the free carrier concentration assumed to be quite negligible compared to N_d. Integrating then from the zero field at the edge of the depletion region ($x = 0$), one finds

$$\varepsilon_0 \varepsilon_r E_p = eN_d d_d \qquad\qquad 4.3a$$

where d_d is the depletion distance on the donor side, with E_p the peak field. Similarly, for the acceptor side

$$\varepsilon_0 \varepsilon_r E_p = eN_a d_a \qquad\qquad 4.3b$$

It can be seen that the total charges on each side, $eN_d d_d$ and $-eN_a d_a$, are equal and opposite, thus keeping the overall electrical neutrality of the junction region. The field in this example falls away linearly to give the total voltage (shaded area in Fig. 4.2d) as

$$\delta V = \tfrac{1}{2} E_p(d_a + d_d) \qquad\qquad 4.4$$

The depletion regions d_a and d_d adjust until this equals the value of δV as given by eqn 4.1.

The position of the conduction or valence band edges can be estimated at any point in the depletion region with respect to the Fermi level by integrating the field up to some more general point x; then $\delta V(x)$ gives the potential at that point and hence the *change* in the position of the band edge (Fig. 4.2b).

If a voltage V_{ar} is now applied in the reverse direction, the charge carriers are pulled further apart and the depletion region widens so as to establish the voltage $\delta V + V_{ar}$. Similarly, if a small forward voltage V_{af} is applied, then the depletion region narrows to establish the potential $\delta V - V_{af}$. This case is indicated in Fig. 4.2d. If the reverse voltage exceeds a value V_p, then the depletion region will run into the n$^+$ contact and, because of the higher donor concentration, $\partial E/\partial x$ will be steeper (Gauss' theorem), giving a field profile as indicated in Fig. 4.2d (dashed line). The depletion region will then widen hardly any further once V_p has been exceeded. The potential V_p is referred to as the 'punchthrough' potential. Before punchthrough is reached eqns 4.3 and 4.4 still hold. In particular, with the example being considered, where $N_a \gg N_d$, one has $d_d \gg d_a$ and we may write the depletion region as merely d, noting that it occurs mainly on the lightly doped side.

Then, for a general voltage V_a (forward direction measured as positive) applied to the terminals of the diode,

$$d \simeq [2(\delta V - V_a)\varepsilon_0\varepsilon_r/eN_d]^{1/2} \qquad 4.5a$$

$$E_p \simeq [2eN_d(\delta V - V_a)/\varepsilon_0\varepsilon_r]^{1/2} \qquad 4.5b$$

One should note that the forward voltage cannot exceed δV or the depletion region will cease to exist. In any case the current flow would become so large that the assumption of negligible charge carriers in the transition region would be unjustified. One should also note that as the reverse voltage is increased (V_a negative) the peak field increases as the depletion region widens. The peak field may not exceed some critical value, determined by the electrical breakdown characteristics of the material, without a significant increase in the current flow compared to the usually negligible reverse value. This breakdown is discussed in greater detail in Chapter 8.

Quasi Fermi levels

It is often convenient to retain the formalism of the Fermi-level approach and draw band diagrams even in the non-equilibrium conditions with an applied voltage. Thus with the Boltzmann approximation one could write $n = N_c \exp -(\mathscr{E}_c - \mathscr{E}_{fn})/kT$ and $p = N_v \exp -(\mathscr{E}_{fp} - \mathscr{E}_v)/kT$ with the energy levels $\mathscr{E}_{fn}$ and $\mathscr{E}_{fp}$ chosen so as to give the correct concentrations. These levels would then be referred to as the *quasi Fermi levels*† for electrons and holes respectively.

It is instructive to consider the qualitative quasi Fermi levels for a forward-biased p-n junction. In the simple theory it is assumed that the Boltzmann approximation is valid even in the presence of current flow on forward bias. The fraction of majority carriers which are then able to surmount a potential barrier V_x is given by $\exp -eV_x/kT$. Thus at the potential $-V_x$ [measured from the position of the equilibrium concentration of electrons, N_d, in the n-region] one will find $N_d \exp (-eV_x/kT)$ electrons. The conduction band in this case has risen by an electron energy eV_x. This means that the quasi Fermi level for the electrons is in line with the Fermi level deep in the n-region where there are equilibrium conditions. At the edge of the depletion region on the p-type side, V_x reaches a value $V_d = \delta V - V_a$. The Boltzmann approximation can also be applied to the holes. Now the holes have to surmount a potential barrier $V_{d-x} = V_d - V_x$ so that $p = N_a \exp -eV_{d-x}/kT$ at the point x where $n = N_d \exp -eV_x/kT$. Thus, with this Boltzmann approximation:

$$pn = N_aN_d \exp e(V_a - \delta V)/kT = n_i^2 \exp eV_a/kT \qquad 4.6$$

† Also called Imrefs: originating as a joke from *Fermi* spelt backwards.

Figure 4.3 indicates the scheme of quasi Fermi levels for the electrons and holes with this Boltzmann approximation, showing that the levels have essentially been split by the applied voltage V_a.

Recombination has, however, so far been ignored. This will try to restore the equilibrium condition $pn = n_i^2$ and so force the quasi Fermi levels together. In Fig. 4.3 it has been assumed that the recombination is only important at the edges of the depletion where there is a strong excess of

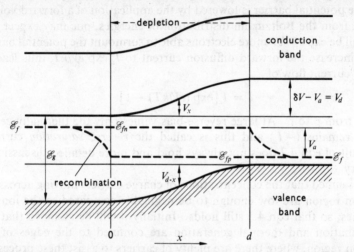

Fig. 4.3 Band diagram for p-n junction with forward bias V_a. δV is diffusion potential; $\mathscr{E}_{fn}$ and $\mathscr{E}_{fp}$ are the quasi-Fermi levels for electrons and holes with the Boltzmann approximation and recombination only at depletion edges. See page

minority carriers, and it is this recombination which brings the levels $\mathscr{E}_{fn}$ and $\mathscr{E}_{fp}$ together. The details of recombination of the minority carriers at the edges of the depletion region are left for discussion in the next section. The band diagram can be similarly drawn for reverse bias, and this is left as an exercise for the reader. The concepts of quasi Fermi levels, though useful for many detailed theories, will not be used extensively in this book.

4.3   The current flow

In order to show the form for the current–voltage relationship we begin by giving a physical argument as follows. The built-in electric field across the p-n junction tends to sweep any minority carriers generated in the p-side edge of the depletion region over to the n-side. Equally, holes generated in the n-side tend to be swept into the p-side. Although the two types of charge carrier are moving in the opposite directions they contribute current in the same sense because of their opposite polarity. Thus thermal generation of minority carriers gives a net current flow of I_s, flowing from n to p (or $-I_s$ flowing from p to n). This current is substantially independent of the applied voltage because it depends only on the thermal generation rates. Clearly at zero bias there must be a balancing current $+I_s$ flowing from p to n to give zero net current. This balancing current is formed by energetic holes from the p-side diffusing over the potential barrier to reach the n-side; equally energetic electrons also contribute by diffusing over to the p-side. If, however, the potential barrier is lowered by the application of a forward voltage V, then, from the Boltzmann distribution of energies, one may expect that there will be exp eV/kT more electrons able to surmount the potential barrier and to increase the forward diffusion current to I_s exp eV/kT, thus leading to a net current flow of

$$I = I_s[\exp (eV/kT) - 1] \qquad\qquad 4.7$$

flowing from p to n. At large reverse-bias values only the thermally gener-current remains $(-I_s)$ and this is called the *reverse saturation* current. The result of eqn 4.7 does not always hold and more detailed discussion is necessary.

It is assumed that the concentrations of charge carriers moving across the depletion region are low enough to be negligible compared to the ionized impurities, so that eqn 4.5 still holds. Initially it will be assumed that the recombination and thermal generation are confined to the edges of the depletion region, where there are plenty of carriers to assist these processes. As indicated above, the application of a forward bias allows energetic majority carriers to diffuse more easily over the potential barrier of the depletion region, and so leads to a concentration of minority carriers in excess of the zero bias thermal equilibrium value: excess holes in the n-region and excess electrons in the p-region. There is a copious supply of such energetic carriers which can cross the depletion region with thermal velocities around 10^5 m/s. If these carriers are not removed fast enough the potential barrier increases, while if they are removed too fast the barrier falls to allow more over. Now the mechanisms for the removal of minority carriers from the edges of the depletion region involve both recombination with the majority carriers and diffusion towards the contacts. This removal is a relatively slow process in most p-n junctions and it is this that limits the current flow.

The physics of minority carrier flow must now be reconsidered. To apply the work of section 3.7 it will be assumed that the minority carrier concentration is much less than that of the majority carrier concentration even on forward-bias conditions where the minority carriers are enhanced. It will then require only a small percentage change of the majority carrier concentration at the edges of the depletion region to neutralize the excess charge of the minority carriers. The movement of the majority carriers is controlled by any electric fields and, because such small movements of the majority carriers are required, the electric fields will remain close to zero throughout the undepleted regions, even in the presence of current flow. The movement of the minority carriers is then described by eqn 3.13, but only with a diffusion component. The majority carriers have their motion controlled by the requirements that the excess minority carriers be neutralized and that the continuity of total current be conserved (see reference 4.8 for a more mathematical discussion of this point). These basic simplifications can now be applied to the p-n junction.

The minority holes in the n-type material have a concentration density $p(x)$ with $x = 0$ defining the edge of the depletion region (Fig. 4.4).

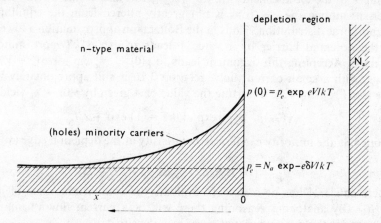

Fig. 4.4 Distribution of minority carriers at edge of depletion region. Small forward bias of V is assumed; p_e is the equilibrium concentration of minority carriers in n-type material; δV is potential barrier required, at zero bias, to align Fermi levels.

Assuming that the majority electrons neutralize these carriers so that the electric field is negligible, the minority carrier current is

$$J_p = -eD_p \, \partial p/\partial x \qquad 4.8a$$

Using the continuity eqn 3.34a

$$D_p \frac{\partial^2 p}{\partial x^2} = \frac{p - p_e}{\tau_{rp}} \qquad 4.8b$$

where the time dependence has been ignored for this steady-state calculation. The solution for this equation can then be written as

$$p - p_e = A \exp{-(x/L_p)} + B \exp{(x/L_p)} \qquad 4.9$$

where $L_p = (D_p \tau_{rp})^{1/2}$.

The parameter L_p is known as the *diffusion length* and the constants A and B must be determined from the physical considerations of the boundaries. Generation and recombination always act so as to make the minority carriers change towards their equilibrium values. If the n^+-n boundary is then ignored, as it is in Fig. 4.4, there can only be a decay of the excess carriers as they move deeper into the majority carrier region away from the depletion edge. Thus in eqn 4.9, $B = 0$.

The constant A is given by evaluating $[p(0) - p_e]$ for arbitrary voltages applied across the diode. The equilibrium concentration p_e is given from the Boltzmann relation (eqn 3.17 with eqn 4.2) as $p_e = N_a \exp(-e\,\delta V/kT)$, where δV is the built-in potential barrier. With a small forward bias the changes in the electric fields are not very great, and so the distribution of energies among the electrons is not greatly altered from the equilibrium Fermi-Dirac distribution. Thus in the Boltzmann approximation a lowering of the potential barrier by a value V leads to $\exp eV/kT$ more minority carriers. Accepting this argument leads to $p(0) = N_a \exp -e(\delta V - V)/kT$. Thus, with a small current flow, relation 3.16 is still approximately valid (see also Problem 4.2). Inserting the value of A given by $p(0) - p_e$ yields

$$p(x) - p_e = p_e[\exp eV/kT - 1] \exp -x/L_p \qquad\qquad 4.10$$

From 4.8a the minority carrier current density at the depletion edge ($x = 0$) is

$$J_p = (D_p/L_p)(en_i^2/N_d)(\exp eV/kT - 1) \qquad\qquad 4.11a$$

where eqn 3.24 has been used to evaluate the equilibrium value of p_e as n_i^2/N_d. By analogous reasoning there will be a current flow of minority electrons in the p-type material given by a density J_n

$$J_n \simeq (eD_n/L_n)(n_i^2/N_a)(\exp eV/kT - 1) \qquad\qquad 4.11b$$

where $L_n = (D_n\tau_{rn})^{1/2}$. These two current densities could be added together for the total current flow in the steady state through the diode. However, given that N_a, the acceptor density, is very much higher than the donor density N_d and that the diffusion lengths are at least comparable, then the current is carried mainly by holes, with eqn 4.11b being negligible because of the large value of N_a. Hence in eqn 4.7 one has

$$I_s \simeq (\text{Area of diode}) \times (en_i^2/N_d)(D_p/L_p) \qquad\qquad 4.12$$

It is emphasized that in most p^+-n diodes the current is carried mainly by the holes and equally in n^+-p diodes the current is carried mainly by the electrons. This will be an important feature when transistors are considered in the next chapter.

The observant reader will possibly be worried at this stage because the diffusion current, evaluated by eqns 4.8 and 4.10, varies with position, thus apparently violating the continuity of current. However in the preamble it was asserted that the electric field in the body of the n-region was *almost* negligible. In the body of the n-region where the minority carriers have decayed and the majority carriers are uniform diffusion cannot play any role. Consequently an electric field must exist, albeit a relatively small one, which forces the continuity of current (by allowing the current to be carried by the majority carriers). Thus it is the majority carrier flow which adjusts the field and takes care of the continuity of current away from the depletion edge. Evaluation of the current at the depletion edge is a simple device for avoiding the mathematical complexities of the interaction between majority and minority carriers.

Deficiencies of the model

If the diffusion length L_p is very long in the n-region or the n^+-contact is fairly close to the depletion region, then the numerical value of the reverse saturation current I_s is influenced. For example, if the contact is at a distance L from the depletion zone, then I_s can become (see Problem 4.3)

$$I_s = \text{Area} \times (eD_p/L_p)(n_i^2/N_d) \coth L/L_p \qquad 4.13$$

for the asymmetric diode. Of course if recombination was too high in the other (p^+) contact, giving a very short L_n, then the contribution to I_s from the minority carrier recombination in that region would not necessarily be negligible and a further increase in I_s would result.

In early p-n junctions the surface leakage was probably the most significant factor to the value of I_s. With modern techniques of manufacture where the surface is protected by an oxide layer this leakage is very much reduced and is usually neglected. Another factor which increases the reverse current is generation in the depletion region. It will be recalled that this was neglected in the simple theory, but this is unsatisfactory for many materials where the minority carrier lifetime is extremely short (e.g. ~nanoseconds for GaAs and gold-doped Si). One finds that one has to add a thermal generation current of order n_i/τ_r per unit volume of depletion region to the value obtained previously for I_s. More detailed estimates can be found in references 4.9 and 8.3.

Recombination in the depletion region tends to restore the equilibrium condition $pn = n_i^2$ in the middle of the region, though with a low value of both p and n. If this happens, one finds that the forward characteristics are affected as well as the reverse current being increased. The equilibrium region in the middle effectively divides the diode into two diodes in series, each carrying the same current and obeying some diode law $I = I_1 \exp eV_1/kT$ and $I = I_2 \exp eV_2/kT$ with $V_1 + V_2 = V$, the total diode voltage; hence

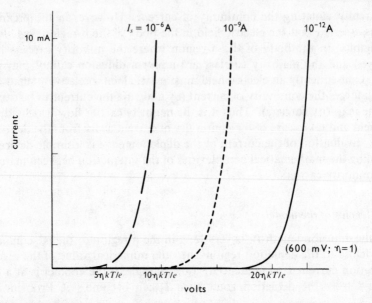

Fig. 4.5 The influence of I_s on diode construction. $I = I_s \exp eV/nkT$. ($kT/e \sim 26$ mV at room temperature.)

$I \sim (I_1 I_2)^{1/2} \exp eV/2kT$. Thus rapid recombination changes the diode law from the ideal relation of eqn 4.7.

The diode law can be changed from the ideal of eqn 4.7 for yet other reasons. The previous work has assumed that the minority carriers have concentrations that are well below the majority carrier levels; this is the case of *low-level injection*. At high-level injection the two carriers approach each other in concentration. If N_n, and P_n are the electron and hole densities on the n-side, then from eqn 4.6: $N_n P_n = n_i^2 \exp eV/kT$. Now as the forward bias V_a approaches δV we may take N_n to be approximately P_n because, although the minority carrier concentration will become very large, the processes of charge neutralization will prevent P_n exceeding N_n, so

$$N_n \sim n_i \exp eV/2kT$$

If one assumes that the current flow is still determined by the minority carrier concentration N_n, then it follows that

$$I \simeq I_s \exp eV/\eta kT \quad \text{with} \quad 1 \lesssim \eta \lesssim 2 \qquad 4.14$$

though I_s may not now be the same as the previously established value. Equation 4.14 will be the most realistic form of diode law in general. Care has to be taken if a diode is to be made that approaches the ideal form of eqn 4.7 over many decades of change in current.

In many applications the alterations to the diode law and reverse saturation current are of little concern, because one is merely using the asymmetric properties of conduction of the p-n junction. The reader should, however, be aware of the differences between the ideal and practical cases. In particular he should have an appreciation of the importance of the saturation current I_s in determining the level of voltage at which the diode starts to conduct. Figure 4.5 shows this effect. It is commonly stated that Si diodes do not start to conduct well until the forward-bias voltage reaches approximately 0·6 volts, but it must be remembered that this is implying current levels in the milliamp range rather than the microamp range which will be reached at a voltage 150 mV or so lower. Since the saturation current depends on the device area, the precise voltage level will also depend on the device area, but any figure is only a rough, though useful, guide.

4.4 The rectifier

One of the simplest applications of the p-n junction is shown in Fig. 4.6, where it is being used to provide a direct current from an alternating source. Since the reverse saturation current is so low one can say that the diode allows current to flow more easily in one direction. Indeed, in the

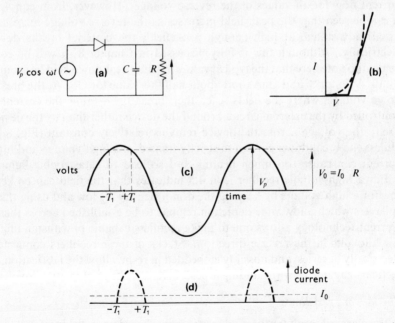

Fig. 4.6 The rectifier or detector. (*a*) Circuit. (*b*) Idealized characteristics for analytical calculation. (*c*) Supply volts. (*d*) Diode current showing conduction angle (see Problem 4.4).

application shown it is usual to consider the diode as almost ideal in that any resistance in the forward direction is taken as constant, and when a reverse voltage appears across the diode it is taken to be open circuited. The solution of this problem is left to the reader (Problem 4.4) where it can be seen that the voltage developed is close to the peak a.c. voltage swing. Such a circuit is used for detecting the envelope of the peak voltage of radio-frequency waves, though in such cases the diode is called a detector rather than a rectifier. From the diode characteristics (Figs. 1.1 and 4.5) it can be seen that forward conduction does not readily occur until some potential ϕ is reached. For silicon p-n junction diodes one loosely states that this potential is 0·6 volts while for Ge p-n diodes it is around 0·3 volts. There are other diodes (see the Schottky barrier diode where $\phi \sim 0.4$ volts for a Si type, and the backward diode in section 4.6 which has a very small effective ϕ) which behave differently and can be much better for high-frequency r.f. detectors. The 'threshold' voltage implies that the diode will be a poor detector for r.f. voltages below ϕ. To ameliorate this defect, one can have a steady bias current to lower the effective value of ϕ, but bias currents larger than a microamp tend to create noise that swamps the very low-level r.f. signals for which one is searching.

Returning to the rectifier, it will be seen that the diode must block the current flow for all values of the reverse voltage. However, from eqn 4.5b it can be seen that the peak field increases as the reverse voltage increases. As already stated, at high enough peak fields the material breaks down electrically. Although this is fully discussed in Chapter 8, it will be convenient to note here that the typical values for this breakdown field lie around $4 . 10^7$ V/m for Si and GaAs and about half this value for Ge. At the breakdown voltage, where the fields reach their breakdown value, the current is controlled by the external circuit around the device rather than by the device itself; the voltage across the device remains relatively constant (Fig. 8.1). Unless the breakdown current is strictly limited, the high voltage and high current can cause too much heating and so lead to catastrophic failure!

For a high-voltage rectifier, eqn 4.5 indicates that the field can be kept low for a high voltage by keeping the donor density N_d low and using thick epilayers which allow wide depletion regions to be established across them. Current technology allows one to make rectifiers capable of withstanding a few kilovolts in the reverse direction. Stacks of these rectifiers connected electrically in series, and possibly embedded in resin, allow the fabrication of rectifiers capable of withstanding a peak inverse voltage of many kilovolts.

The forward resistance

The classical result for the diode resistance is based upon the ideal voltage–current characteristic given in eqn 4.7. The dynamic conductance for the diode is given by $g_d = \mathrm{d}I/\mathrm{d}V$ so, from eqn 4.7, $g_d = [I_s(kT/e)]\exp eV/kT$—or allowing more generally for non-ideal behaviour (as in eqn 4.14)

$$g_d \sim I/(\eta kT/e) \qquad (1 \leqslant \eta \leqslant 2) \qquad 4.15$$

This value is useful for considering relatively small changes in current and voltage about some steady-state condition. Such *small signal calculations* allow a linear theory to be used to describe changes of current and voltage about the bias conditions. Thus in eqn 4.15, the steady bias current I should be used and a common figure for the forward resistance is given by $r_e = 1/g_d \simeq (25/I)$ ohms, where I is in milliamps and room temperature operation is assumed. For a rectifier one might take the forward resistance to be given with half the peak current as the value for I, but of course on reverse bias the resistance would become very high.

The resistance calculated above is independent of the area of the diode, depending only on the level of current flow. However, the current has to flow through metal contacts and through the passive parts of the semiconductor before reaching the depletion region. This contributes a resistance r_b in series with the ideal diode resistance r_e. The value of r_b varies inversely with the area and depends on the technology that is used. Typical values of r_b for the construction outlined in section 4.1 are around 1 ohm for 10^{-8} m^2 of device area. At high current levels it is this resistance which is important rather than r_e.

At high current densities there are additional causes of voltage drop which add to the resistance, and also change the I–V characteristics. The charge carriers create their own electric fields and these fields change the voltage (usually increase the forward bias voltage). Eventually the current flow becomes limited by these space charge fields (*space charge limited flow*), but this is left for further reading.[1.5]

4.5 Junction and diffusion capacitances

The rectifier or detector fails to work effectively at high enough frequencies because charge is stored in the p-n junction and the charge cannot be changed sufficiently rapidly. There are two forms for the charge storage: charge of the fixed ionized donors in the depletion region and charge of the excess minority carriers stored at the edge of the depletion region. The former leads to a *depletion capacitance*, the latter to a *diffusion capacitance*.

The depletion capacitance is obtained by noting that the charge stored on each side of the depletion region is given from Gauss' theorem as

$$Q = \varepsilon_0 \varepsilon_r E_p/\text{unit area} \qquad 4.16$$

From eqn 4.5*b*, with an applied voltage V

$$Q = [2eN_d\varepsilon_0\varepsilon_r(\delta V - V)]^{1/2} \qquad 4.17$$

The depletion capacitance is defined as dQ/dV, a differential capacitance relating small changes of charge and voltage. This is useful for the same reasons (the linear analysis of non-linear problems) that lead to the dynamic conductance g_d (eqn 4.15). Thus differentiating eqn 4.17:

$$C_{dep} = dQ/dV = \tfrac{1}{2}[2eN_d\varepsilon_0\varepsilon_r/(\delta V - V)]^{1/2} \quad \text{per unit area}$$

$$= \varepsilon_0\varepsilon_r/d \quad \text{per unit area.} \qquad 4.18$$

The result of eqn 4.18, obtained using eqn 4.5*a*, is a universal result regardless of the particular impurity profile at the junction. The charge only changes at the edges of the depletion region, so that for small signals the depletion layer behaves as a parallel-plate capacitor of width d, the depletion width.

The diffusion capacitance is more complicated and only a simple presentation is given here, based on the concepts of 'charge' control of the current by charge storage of the minority carriers. In the model for the current flow that gave rise to eqn 4.11*a* one could integrate the excess minority carriers, given by eqn 4.10, to find the total charge stored by the diffusion process as

$$Q_s = ep_eL_p(\exp eV/kT - 1) \times \text{area}$$

From eqn 4.11*a* and substituting for L_p from the value given in eqn 4.9, one finds

$$Q_s = I \cdot \tau_r \qquad 4.19$$

The subscript p has been dropped on the recombination time to indicate the wider generality of the result than our specific case. The current is then seen to be controlled by the amount of stored minority-carrier charge in excess of the equilibrium amount. This then is the concept of charge control: control the charge and hence the current. The excess minority carriers are in turn controlled by the applied voltage through the control of the barrier height for the excess flow. Now to reduce the charge by δQ_s in a time δt requires a decrease in the current by an amount $k\,\delta Q_s/\delta t$. The constant k is usually less than unity because the current takes time to distribute the carriers with the correct spatial distribution (Fig. 4.5). This allows recombination to assist in the reduction of the stored charge. Thus, in general, not all the stored excess minority charge can be recovered; recombination helps to remove some of it. A fuller discussion is given in reference 1.2. In our example $k = \tfrac{1}{2}$ is found to be the appropriate value from this more detailed work. The total current flow at any stage is then

$$I = k\frac{dQ_s}{dt} + \frac{Q_s}{\tau_r} = C_{dif}\frac{dV}{dt} + \frac{Q_s}{\tau_r} \qquad 4.20$$

where $$C_{\text{dif}} = k \ \mathrm{d}Q_s/\mathrm{d}V = k\tau_r(\mathrm{d}I/\mathrm{d}V) = k\tau_r g_d \qquad 4.21$$

The dynamic conductance g_d should not now be taken, as in eqn 4.15, to be a function of the current flow, but to be a function of the stored charge Q_s—as an approximation $g_d \sim eQ_s/(\eta\tau_r kT)$. One notes that the diffusion capacitance is not a function of the area but only of the stored charge (or current for the *steady*-state case). Figure 4.7 then gives a useful first-order equivalent circuit based on these small signal parameters. On reverse bias, g_d and C_{dif} are negligible. In practical calculations, when currents are switched one must allow for the non-linear nature of these circuit parameters. This is emphasized in the next section.

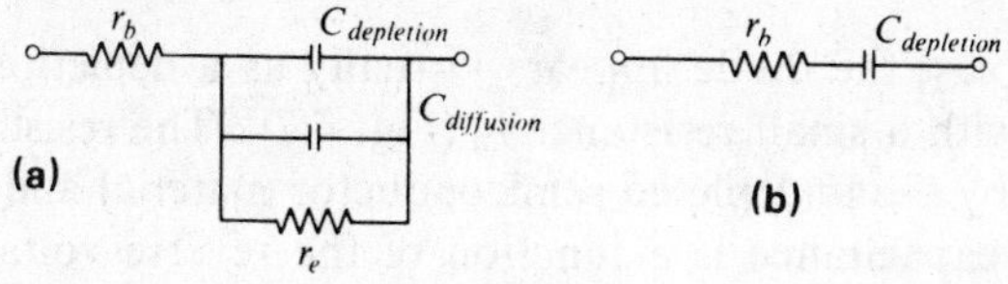

Fig. 4.7 Equivalent circuit for p-n junction. (*a*) Forward bias. (*b*) Reverse bias (I_s negligible). Note that $C_{\text{depletion}}$ is not the same in (*a*) and (*b*) because of bias voltage change.

4.6   Types and uses of junction diodes

The fast switching diode

In many applications the diode is used as a simple switch: under forward currents its impedance is low, while on reverse bias its impedance is high. Suppose that the current is changed abruptly from zero to a forward value of a few milliamps; the conductance and capacitance then change from the low values at reverse bias to the higher values at forward bias. However the higher values can be achieved only after the minority carriers have rearranged themselves; initially when the current changes abruptly the impedance remains high and the voltage rises well above the equilibrium value and then subsequently decays, with a time constant given by the recombination time τ_r, to the value appropriate for the particular forward current. Similarly, when switching back from a forward to a reverse state, the diode initially remains in a low-impedance state until the minority carriers have been removed—typically again with a time constant τ_r,—when the high-impedance state is achieved. Clearly it is important to use the non-linear nature of g_d and C_{dif} in Fig. 4.7 to understand the effects just described.

To make the diode switch very rapidly from the off- to the on-state demands a short recombination time in the p-n junction. This is achieved by adding impurities which speed up the recombination. For example, gold is an interstitial impurity in Si which happens to give both a donor and acceptor energy level close to the middle of the band gap. This tends to compensate the Si (section 2) and, more importantly here, to speed up recombination (see Appendix 4). It is found that gold-doped Si diodes can have switching times in the nanosecond range. This is a common solution for the manufacture of fast switching diodes for computer applications. In Chapter 7 a different type of diode will be discussed—the Schottky barrier diode— where minority-carrier storage problems are almost eliminated and the diffusion capacitance is negligible. Switching times in the picosecond range are then possible.

The varactor (variable reactor)

On reverse bias, the diode appears virtually as a depletion capacitance C_{dep} in series with a small resistance r_b (Fig. 4.7). The resistance r_b is primarily caused by the undepleted semiconductor material and the contacts. The depletion capacitance is a function of the reverse voltage (eqn 4.18) until punchthrough, when the depletion layer tends to remain at the epilayer thickness (Fig. 4.2d). Thus one has an electronically variable capacitance with the main loss of power coming from the series resistance. A measure of the highest frequency at which this capacitance is useful is given by the cut-off frequency $\omega_c = 1/C_{dep}r_b$. Because both the capacitance and resistance are functions of applied bias it is usual to specify the cut-off frequency at a fixed bias (often zero volts). Varactors with values of ω_c around several hundred GHz have been made using GaAs.

One of the more interesting uses of the varactor comes in a device called the Parametric Amplifier.[4,5] We present here a simplified account which brings out a few of the features of how power can be given out from a time-varying capacitance. The essential features of the circuit requirements are two circuits, one supporting a signal frequency ω_s and the other an idler frequency ω_i (Fig. 4.8) These are connected by a capacitance which is assumed to be modulated by a pump frequency $(\omega_s + \omega_i) = \omega_p$ as $C = C_0/[1 + m \sin(\omega_p t + \psi)]$. In principle, this variation is brought about by the change of voltage across the diode caused by an r.f. supply at the pump frequency. Provided that ω_p is well below ω_c, the capacitance will be able to respond to the r.f. pump voltage. The charge on the capacitor coupling the circuits is assumed to vary as $q = q_1[\cos \omega_s t + \cos(\omega_i t + \psi)]$. The voltage across the capacitor is given by $V = q/C$. This voltage is split into the component V_s varying as the frequency ω_s and the component V_i varying as ω_i. There is also the pump voltage V_p but this is much larger than the signal or idler voltages and it is this component which is changing the value of the capacitance. The other frequencies in the voltage are ignored, since they will not contribute to the action of the circuit at ω_i and ω_s. From

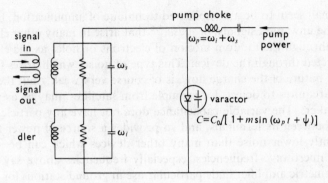

Fig. 4.8 Schematic parametric amplifier circuit. Signal and idler circuits assumed to have low impedance at all frequencies other than their resonant frequencies. Pump choke assumed to have high impedance except at pump frequency $\omega_p = \omega_i + \omega_s$. Note circuit symbol for varactor.

simple manipulation of q/C

$$V_s = (q_1/C_0)(\cos \omega_s t + \tfrac{1}{2}m \sin \omega_s t)$$

$$V_i = (q_1/C_0)(\cos (\omega_i t + \psi) + \tfrac{1}{2}m \sin (\omega_i t + \psi)$$

The power at each frequency is found from the average value of $V(dq/dt)$; thus $P_s = \omega_s(mq_1{}^2/4C_0)$; $P_i = \omega_i(mq_1{}^2/4C_0)$ gives the power at the signal and idler frequencies respectively, the sense of the power flow being outwards from the variable capacitor. One finally notes, from the conservation of power, that the pump power $P_p = P_i + P_s$ satisfies the relationships:

$$P_p/\omega_p = P_i/\omega_i = P_s/\omega_s \qquad 4.22$$

These are simple examples of more general power relationships which hold for non-linear reactances and which were first worked out by Manley and Rowe.[4.7]

The idler circuit has the important function of adjusting its phase to suit that of the pump, so that power can be given out. This adjustment comes about automatically in practice, although in the mathematics the two phase factors for the pump and idler have been both put equal to ψ as one of the assumptions. It is possible to make a *degenerate* parametric amplifier— one where the signal and idler have degenerated to identical frequencies— but then it is found that the phase of the pump power is critical in relation to the signal phase if power is to be given out. Degenerate amplifiers are rarely used. In summary, the parametric amplifier gives out power at the signal frequency by pumping a capacitor with some different but higher frequency. A small amount of power coupled into the signal circuit can then permit larger amounts to be coupled out.

This may seem to be a complicated technique of amplification, but it removes the unwanted signals, or *noise*,[4.6] that arise in many other electronic devices through the random motion of electrons or holes as the electricity is conducted through the device. This type of noise, which arises from the granular nature of the charge flow, is of course very weak,† but the signals that one requires to detect, for example from satellites, may be as weak or even weaker. The variable capacitance does not have any particle current flowing between its terminals, and so provides a source of power that has significantly lower noise than many other devices which can be made to work at microwave frequencies, especially frequencies above say 5 GHz. The parametric amplifier finds particular use in ground stations for satellite communications.

The tunnel diode[4.2]

According to classical mechanics, an electron with an energy $\mathscr{E}$ cannot surmount a barrier with an energy greater than $\mathscr{E}$, no matter how thin the barrier. However, quantum mechanics shows that the quantum electron wave does have a finite probability of penetrating a barrier. In Problem 6.2, using standard wave mechanics, it is shown that the probability of an electron 'tunnelling' through a barrier of uniform potential V_b, in excess of the kinetic energy, and of thickness L is the order of $\exp -L(8m^*eV_b)^{1/2}/\hbar$. Remembering that there can be very large concentrations of charge carriers in the valence band of semiconductors, the reader should convince himself (by inserting some orders of magnitude) that significant concentrations of charge carriers could cross a barrier about 1 eV high and 10 nm thick. This process has direct application to a p-n junction. A semiconductor in an electric field E has the valence band and conduction bands at the same level of energy with a spatial separation of $L = V_g/E$ where eV_g is the band-gap energy. Consequently one may expect a tunnel current from the valence band into the conduction band which will vary as $I = A \exp \{ -B(m^*e)^{1/2}V_g^{3/2}/\hbar \}$. The factor B is a number of order unity and depends on the detailed shape of the potential barrier between the conduction and valence band. The factor A depends on the number of quantum states available to give electrons and to receive electrons in the tunnelling process. From the numbers given, it is apparent that electric fields in the range of 10^8 V/m are required. Such high fields can be obtained in a semiconductor at relatively low voltages only if the doping density is very high (see for example eqn 4.5b). Indeed, by making diodes with very high impurity concentrations (p^+-n^+ diodes) a new type of diode characteristic is observed: the *tunnel diode*. To make such a diode the Fermi levels have to lie inside the bands rather than in the

† Around 10^{-14} watts for a device with 1 MHz bandwidth.

band gap (Fig. 4.9*a*). With high enough impurity concentrations, the fields in the depletion region reach the 10^8 V/m range even at zero bias where the Fermi levels are aligned. Provided that there are states to supply electrons and to receive the electrons at the appropriate energy, tunnelling can occur

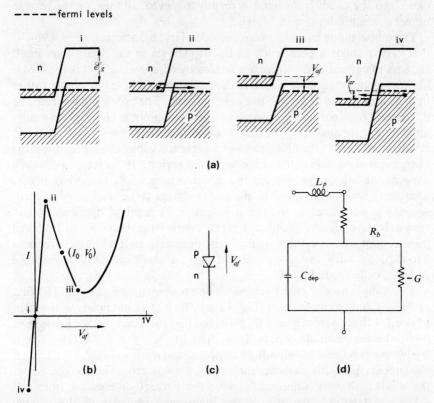

Fig. 4.9 The tunnel diode. (*a*) Band diagrams at different applied voltages. (*b*) *I–V* characteristics. Voltage scale set by band gaps and Fermi levels; current scale set by device area; typical peak currents about 1–5 mA. (*c*) Circuit symbol. (*d*) Small-signal equivalent circuit about (I_0, V_0). The valley voltage for Ge diode is about 0·35 V, for Si about 0·5 V, and for GaAs about 0·7 V.

across the band gap—though at zero bias the tunnel currents in either direction must cancel for equilibrium (point i on characteristic in Fig. 4.9*b*). As the forward bias is increased by a few kT/e volts, the edge of the conduction band will start to move above the edge of the valence band and the tunnel current will decrease (Fig. 4.9*a* iii) or even cease as the voltage is raised still further. This happens because there are no longer permitted energy states in the valence band opposite the conduction band for the transition of electrons (through the potential barrier) to occur. The current then falls to the level of the diffusion current, i.e. that level determined by the diffusion of *energetic* carriers *over* the potential barrier as in normal diode conduction. As the voltage is raised still further the normal diode characteristics are followed. On reverse bias the tunnel current of electrons from the valence band into the conduction band is readily achieved, so that a large reverse tunnel current flows (Fig. 4.9a, iv).

The tunnel diode has many uses because of its characteristic (Fig. 4.9b). If it is biased about a point such as (I_0, V_0) shown in the figure then small changes of current and voltage are related by $dI/dV = -G$; this is called a negative conductance. A negative conductance can give out power rather than absorb power like a positive conductance. The power comes from the d.c. supply required to keep the mean operating point at (I_0, V_0). An equivalent circuit for such changes can be given for the tunnel diode (Fig. 4.9d). It can be noted that the diode still has a depletion capacitance because of the charge stored on either side of the depletion region. However, the diffusion capacitance can be ignored for the active region where tunnelling occurs because the tunnelling process allows the charge to be readjusted in picoseconds as compared to the much longer times required by diffusion and recombination. There is still the series contact resistance R_b to be taken into account and, as with any really high-speed devices, it is as well to remember that the leads to the device contribute a small but finite inductance L_p—typically around 1 nH.

Two simple uses[4.2] of the tunnel diode are shown in Fig. 4.10. The first of these is that of an oscillator (Fig. 4.10a) which gives out power in the form of an r.f. voltage across a load G_0 provided that this conductance is less than $|-G|$, the tunnel diode conductance. The frequency of oscillation is controlled by the resonant circuit at approximately $\omega^2 = 1/(C_{dep} + C_0)L_0$ if the package parasitic elements such as L_p are ignored. The voltage across the diode will vary sinusoidally with the current determined from the I-V characteristics. Because of the high-speed response of the voltage across the tunnel diode to changes in current, such oscillators can be made to work at frequencies well above 10 GHz, though special circuits have to be employed and special diodes with well-designed packages must be used to keep the stray capacitances and inductances low.

The second use (Fig. 4.10c) of the tunnel diode is as a simple switch. The diode is normally biased at a point A on the I-V characteristic (Fig. 4.10d) by a supply V_{bb} and a load resistor R. One notes that the device voltage V and current I must satisfy the *load line*, $V = V_{bb} - IR$, a straight line on the I-V graph. The load line cuts the diode characteristic at two more points. The middle point is unstable and the device cannot sit at that point. Thus if an input voltage takes the current momentarily over the peak value I_p, the diode switches to the point B at a higher voltage and lower current.

It will remain there until the input voltage brings the current down to below I_v, when it will switch back to the high-voltage state. A tunnel diode can switch in picoseconds, and pulse generators with 50 picosecond pulse widths can be made using these devices.

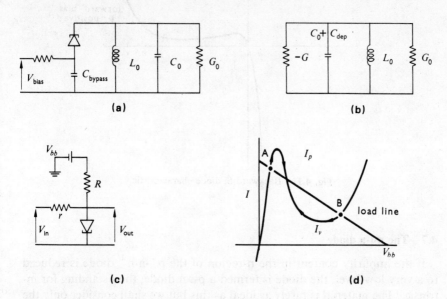

Fig. 4.10 Simple tunnel diode circuits. (*a*) Oscillator. (*b*) Equivalent circuit ignoring L_p and R_b. (*c*) Switch circuit. (*d*) Load line showing switching between A and B.

The backward diode

In section 4.4 it was seen that the conventional p-n junction was poor at detecting r.f. voltages below the voltage ϕ required to achieve good conduction. The backward diode is a modification of the tunnel diode which removes both this difficulty and that of minority-carrier storage limiting the frequency response. The required modification is the reduction of the carrier concentration in the p-type material so that the Fermi level lies close to the edge of the valence band in that material. Tunnelling is thus almost prevented on forward bias but is possible on reverse. This gives the diode a characteristic shown in Fig. 4.11. With silicon, it makes a good low-level voltage detector for voltages below about 0·6 volts, but of course produces a detected voltage in the opposite polarity to that of the conventional diode, conducting best on 'reverse' bias with the n-side positive.

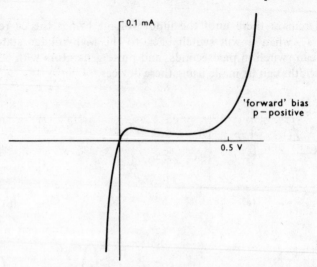

Fig. 4.11 Backward Si diode characteristic.

4.7 The p-i-n diode

If the impurity content in the n-region of the p^+-n-n^+ diode is reduced
to a very low level, the diode is termed a p-i-n diode, the i standing for intrinsic. The material is rarely as ideal as this but we shall consider only the
ideal case. The band diagram at equilibrium is shown in Fig. 4.12a. On
reverse bias, the depletion region extends right across the intrinsic region
into the edge of the p and n contacts (Fig. 4.12b). Because there are no
impurities in the i-region there is no charge from the ionized impurities and
so the field is constant ($dE/dx = 0$ from Gauss). The field only falls to zero
at the highly doped contact. By making the width of the i-region long, the
breakdown field E_b can be reached only at large voltages $V_b = E_b \cdot d$. Thus
p-i-n diodes make good high-voltage rectifiers with i-regions in the 100–
1000 μm range. The useful limit to the length is set by the lowest impurity
content available rather than the width of the i-region (see eqn 4.5 for the
peak field).

The p-i-n diode can also be used to control microwaves. With a large
i-region the capacitance on reverse bias can be very low. This capacitance
is given approximately at all reverse bias voltages by $C_{dep} = \text{Area} \cdot \varepsilon_0\varepsilon_r/d$.
Thus, by choice of d, this can be made to be a high impedance even at micro-
wave frequencies ($\sim 10\,\text{GHz}$). On forward bias, however, the i-region
becomes filled with charge carriers injected from the p and n regions. The
impedance to high frequencies then becomes low and resistive. Thus by

changing the d.c. bias through a p-i-n diode its impedance to microwaves can be changed; this is one special use for this device.

The p-i-n diode does not behave in quite the same manner on forward bias as the conventional p-n junction. One should note from the band diagram that the potential of the conduction and valence bands changes markedly at the edges of the i-region in order to align the Fermi levels correctly. The sense of these resulting fields is in the direction to prevent too many majority carriers diffusing out from their respective regions. It follows then that any holes close to the n-region are pushed back into the i-region and electrons near the p-region are pushed back into the i-region. Thus injected carriers are confined mainly to the i-region by these fields. Figure 4.12c shows the charge and field distributions on forward bias, at least schematically. The precise distribution of the injected charge carriers is not important here; it is merely a more or less uniform distribution of equal numbers of holes and electrons—the holes with a charge Q_s. With a recombination time τ_r, the charge control equation is:

$$I = \frac{dQ_s}{dt} + \frac{Q_s}{\tau_r} \qquad 4.23$$

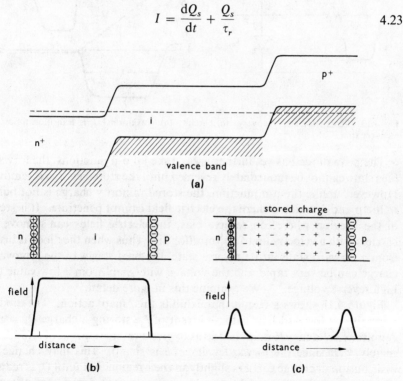

Fig. 4.12 The p-i-n diode. (a) Band diagram at zero bias. (b) Field and charge on reverse bias. ⊕⊖ fixed ion charges; ± mobile charges. (c) Field and charge on forward bias.

for forward bias. In the steady state, $I\tau_r = Q_s$. The voltage applied across the terminals to obtain a current I will follow a law similar to that of eqn 4.14, with $\eta \sim 2$. Thus the diode will have a low voltage and a high charge if τ_r is large enough, and consequently will look like a high capacitance on forward bias. However, on reverse bias, once the charge carriers have been removed from the depletion region, the diode will look like a low capacitance. This *varactor-like* behaviour is found to be very useful for non-linear applications where signals of different frequencies have to be mixed together. It is not used for parametric amplifiers because, to obtain the very large change of capacitance, one has to forward-bias the diode. The forward current flow would add noise to the signal.

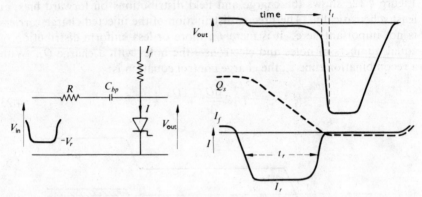

Fig. 4.13 Circuit for step recovery. (*a*) Circuit. (*b*) Waveforms: t_t is transition time; t_r is recovery time.

The p-i-n diode has yet further uses. Like all p-n junctions, the forward-bias state cannot be immediately reversed until the stored charge is removed. However, unlike the p-n junction, the stored minority charge is not buried in the p and n regions where the electric field cannot penetrate. It is stored in the i-region where, on reverse bias, the electric fields can remove the carriers with a velocity up to 10^5 m/s (for Si). Thus when the diode is biased from a forward state into a reverse state, the final stages of the removal of charge can be very rapid and the voltage will 'snap' from a low value to a high reverse voltage.[4.3] We examine this in more detail.

Figure 4.13 shows a circuit that exhibits this 'snap' action. The diode is normally on forward bias with a current of I_f, storing a charge $Q_s = \tau_r I_f$. A pulse of voltage $-V_r$ is applied so as to give a reverse current of I_r (approximately V_r/R since the forward voltage ϕ is small). This may change the distribution of charge carriers slightly in the i-region, but until Q_s is reduced nearly to zero the edges of the p and n regions cannot deplete and allow the full reverse voltage to be established. Thus the large amount of stored charge keeps the voltage at a low value during extraction of the stored charge. This lasts for a recovery time t_r and it is left to the reader to show from eqn 4.23 that $t_r = \tau_r \log_e [1 + (I_f/I_r)]$, at which time $Q_s = 0$ according to this simple model. The diode can then change from a very high capacitance to its low reverse-bias capacitance C_{dep} and the voltage V can snap back to $-V_r$ with a time constant RC_{dep}. Usually this time constant is so low that it is the velocity at which the carriers can be removed which limits the snap action rather than the RC time constant. This fundamental limit gives a 'snap' time-constant of approximately 10 picoseconds per micrometre length of i-region. The idealized waveforms are indicated in Fig. 4.13b, showing how the stored charge Q_s decreases and the snap action occurs. More detailed analysis shows that the transition to the high voltage has two phases, a slow phase and the fast transition mentioned here. One can see that to utilize this fast step of voltage it will pay, in contrast to the high-voltage rectifier, to keep this i-region short—typically in the range $\frac{1}{2}$ to 5 μm. Such diodes are known as *step-recovery* diodes or *snap* diodes.

The step-recovery diode (SRD) has several uses. It can be seen that, to give the sharp transition, one does not first require a fast rising pulse. Thus the leading edge of a pulse can be sharpened from say a nanosecond to 50 picoseconds rise time by the use of a SRD. The 'pulses' used to trigger the reverse-bias state can in fact be formed by a sine wave, so that a train of fast rising pulses is generated by the input sine wave. Such a train of pulses is rich in harmonic content. One can produce harmonics very efficiently and so generate higher frequencies from a given source. This technique is called *frequency multiplication* and the SRD is a commonly used component.

The SRD is a useful note on which to end this chapter. It draws attention to some of the special properties of p-n junctions in a very marked manner. If one merely wants rectification or variable low-loss capacitances one will find that metal–semiconductor junctions (Chapter 7) have similar properties but without the charge-storage properties that may be detrimental to these straightforward applications.

PROBLEMS

4.1 An impurity concentration in the neighbourhood of a plane $x = 0$ varies as $N_i(x/L)$, a positive sign being taken for say the positive charge of the ionized acceptors and a negative sign for the negative sign of the ionized donors. Assume that the depletion extends from $+d$ to $-d$ and that the Debye length at these points is short compared to d, so that the abrupt depletion zone can be assumed. Show that the built-in potential at equilibrium is given by $\delta V = 2N_i d^3/3\varepsilon_0\varepsilon_r L$ where $n_i \exp e\, \delta V/2kT = N_i(d/L)$. Show also that at large reverse-bias voltages the differential capacitance dq/dV is proportional to $V^{-1/3}$.

4.2 Show that for a constant current J the equation

$$J = ep\mu E - eD\frac{\partial p}{\partial x}$$

can be written as

$$J \exp(eV/kT) = -D \frac{\partial}{\partial x}(ep \exp eV/kT)$$

where V is the potential of the valence band. Hence integrate this equation to show that across a p^+-n junction the hole current is

$$J = DN_a \exp -e\, \delta V/kT[\exp(eV/kT) - 1] \Big/ \int_{-(\delta V - V)}^{0} \exp(eV(x)/kT)\, dx$$

where the p^+ layer has an acceptor concentration of N_a (potential $-\delta V + V$ on forward bias) and the n-layer has an equilibrium concentration of $N_a \exp -e\, \delta V/kT$ (taking the reference potential of 0 for the valence band). Note the difficulty in evaluating this integral with any accuracy because of the exponential denominator.

4.3 In a certain diode the contact to the n-region is only a short distance L from the depletion edge. At this contact the recombination is so rapid that the minority hole concentration can be taken as the equilibrium value. With this assumption, evaluate the constant B in eqn 4.9 and, using eqn 4.8a evaluated at the depletion edge, prove the result of eqn 4.13.

4.4 A rectifier starts to rectify at a small potential ϕ, when it can be regarded as a constant resistance R_b. For reverse bias or voltage less than ϕ the device can be considered to be open circuit. From Fig. 4.6, with the assumption that C is large enough to make the voltage across R a constant V_0, show that for

$$(1/2\pi)\int_{-\theta_1}^{+\theta_1}(V_p \cos\theta - V_0 - \phi)\, d\theta = (R_b V_0/R)$$

where $\theta_1 = \omega T_1$. From $V_p \cos\theta_1 = V_0$ show that for $\phi \ll V_0$,

$$V_0 \simeq V_p - \tfrac{1}{2}V_p(3\pi R_b/R)^{2/3} \qquad [\tan\theta \simeq \theta + \tfrac{1}{3}\theta^3]$$

4.5 At the edge of a p-i-n diode the n-concentration may be taken as $N_d \exp \beta x$. By putting the electron majority current equal to zero show that in this region there is a built-in field $E = -(kT/e)\beta$. Hence show for the minority excess carriers

$$\frac{\partial}{\partial x}\left(-\beta D p_1 - D\frac{\partial p_1}{\partial x}\right) = -\frac{p_1}{\tau_{rp}}$$

Hence show that the minority carriers decay as $\exp -\beta x$ going into the n^+-region but only decay as $\exp(x/\beta L_p^2)$ going away from the region (x negative). It may be assumed that $\beta^2 L_p^2 \gg 1$ with $L_p^2 = D\tau_{rp}$. Discuss why this shows that the minority carriers are confined to the i-region in the p-i-n diode.

4.6 If in eqn 4.8b time variation is allowed, show that

$$\frac{\partial p}{\partial t} - D\frac{\partial^2 p}{\partial x^2} = -\frac{(p - \rho_e)}{\tau_r}$$

Hence show that the minority carriers vary as $\exp -[x(1 + \tfrac{1}{2}j\omega\tau_r)/L_p] \cdot \exp j\omega t$ for $\omega\tau_r \ll 1$. The excess carrier concentration at the depletion edge is $p_0 + p_1$ where p_0 is the mean value and p_1 is the component varying as $\exp j\omega t$. From the Boltzmann relationship show that $p_1 \simeq (ep_0/kT)V_1$ where V_1 is the alternating voltage about a forward-bias point. From the steady-state relationships show $J_0 = (eDp_0/L_p) \times$ Area while from the alternating relationships show $I_1 = eDp_1[(1 + \tfrac{1}{2}j\omega\tau_r)/L_p] \times$ Area. Hence show that the differential diffusion capacitance defined in eqn 4.21 is $\tfrac{1}{2}\tau_r g_d$, demonstrating that $k = \tfrac{1}{2}$ is appropriate.

General references

SZE, S. M. Reference 1.5.
LINDMAYER, J. and WRIGLEY, C. Y. Reference 1.2.

Special references

Diodes

4.1 WATSON, H. A. *Microwave Semiconductor Devices and their Circuit Applications*. McGraw-Hill, 1969.
Chapters on varactor diodes, p-i-n diodes and tunnel diodes by specialist authors.
4.2 GENTILE, S. P. *Basic Theory and Applications of Tunnel Diodes*. Van Nostrand, 1962.
4.3 MOLL, J. L. and HAMILTON, S. Physical modelling of the step recovery diode. *Proc. IEEE*, (1969), 1250–9.
4.4 SHURMER, H. V. *Microwave Semiconductor Devices*. Pitman, 1971.

Parametric amplifiers and noise

4.5 HOWSON, D. P. and SMITH, R. B. *Parametric Amplifiers*. McGraw-Hill, 1970.
4.6 VAN DER ZIEL, A. *Noise, Sources, Characterisation and Measurement*. Prentice-Hall, 1970.
4.7 MANLEY, J. M. and ROWE, H. E. Some general properties of non-linear elements. *Proc. IRE*, **44** (1956), 904–913.
See also: A simplified derivation of the Manley and Rowe power relationships, by J. E. Carroll. *J. Electronics and Control* **6**, (1959) 359–361.

Recombination and leakage currents

4.8 JONSCHER, A. K. *Principles of Semiconductor Device Operation*. Bell, 1960.
See sections on ambipolar diffusion.
MOLL, J. L. Reference 8.3.
4.9 SAH, C. T., NOYCE, R. N., and SHOCKLEY, W. Carrier generation and recombination in p-n junctions and p-n junction characteristics. *Proc. IRE*, **45** (1957), 1228–1243.

5

Amplifiers and Control Devices using p-n Junctions

5.1 Amplifiers

Amplifiers and control devices accept small changes of current or voltage and convert them into larger changes. An amplifier, as opposed to a transformer, gives out more power in the larger changes than it accepts at the input, the extra power coming from the battery or power supply that must be connected to activate the amplifier. Circuit designs which involve amplifiers are usually based upon the use of equivalent circuits; these relate the changes at the input with those at the output. Figure 5.1 shows such an equivalent circuit with the changes in voltage and current related by a hybrid of input and output quantities as follows:

$$\delta V_{\text{in}} = h_i \, \delta I_{\text{in}} + h_r \, \delta V_{\text{out}}$$
$$\delta I_{\text{out}} = h_f \, \delta I_{\text{in}} + h_o \, \delta V_{\text{out}}$$

5.1

These are the *hybrid* parameters or *h*-parameters for the device in question. It should be noted that they omit any physical mechanism that may be involved inside the device and do not show how the d.c. supplies have to be connected. In general they will be limited in validity, and the parameters will be functions of the bias supplies to the device. In spite of these obvious limitations, they are useful because they help to analyse non-linear devices in terms of linear equation—valid at least for small changes of signal as discussed in Chapter 4. There are other methods of relating the input to the output; the admittance parameters, for example, have the current on one side and the voltages on the other (see Problems 5.7 and 5.9).[1.1]

To see the physical significance of the *h*-parameters, first keep the output voltage constant ($\delta V_{\text{out}} = 0$). The *input impedance* is then seen to be h_i (ohms) while the current at the output is increased over that at the input by the *forward transfer ratio*, h_f. Now allow the voltage to vary but keep the input current constant ($\delta I_{\text{in}} = 0$); the *output conductance* across the output terminals is then h_0 (siemens) while a voltage is transferred back in the reverse direction with a *reverse feedback ratio* h_r. In most good amplifiers, h_r is negligible.

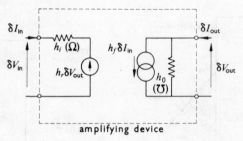

Fig. 5.1   The hybrid-parameter equivalent circuit.

The h-parameters do not give the best representation for all amplifiers but they are well suited to the *bipolar transistor*, as the n-p-n or p-n-p structures are called. These three-layer devices were the first really practical solid-state amplifiers, and so we shall start our discussion on amplifier devices by looking at how they operate. The aim will be to find their h-parameters and then to design a simple amplifier. The action of such transistors will be presented in terms of an idealized model for the n-p-n device which takes advantage of modern construction techniques. The whole set of arguments could of course be rephrased in terms of a complementary p-n-p structure, but this is left as an exercise for the reader.

5.2   An idealized n-p-n transistor

Figure 5.2*a* shows schematically the impurity concentration as a function of distance for an idealized silicon transistor. The left-hand region is the *emitter*. It is very highly doped ($\sim 10^{26}$ donors/m^3); its width is not very important but is typically a few micrometres, or less. This forms a junction with a p-type region called the *base* in which the acceptor impurities fall from around $5 \cdot 10^{23}$/m^3 down to $5 \cdot 10^{20}$/m^3 with an exponential profile. The base width is typically less than a micrometre in this model. The final region is the n-type *collector* which forms another junction on the other side of the base. The collector is ideally made up of two stages: a lightly doped active region which supports a depletion region, and a heavily doped contact or substrate. The reasons for this substrate are essentially those for a substrate to a diode; it gives a low-resistance electrical contact but simultaneously provides a good backing of semiconductor material which eases the difficulties of handling very thin slices.

Figure 5.2*b* shows the circuit symbol for this n-p-n device and it should be noted that the arrow indicates the sense of *conventional* current flow as being out from the emitter, so in this case *electrons* are travelling *from* the emitter *towards* the collector. The symbol for the p-n-p device has the arrow at the emitter pointing the other way, so that the polarity of the device is

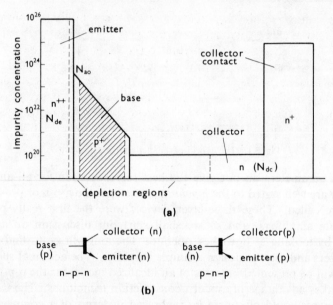

Fig. 5.2 The transistor. (*a*) An idealized structure. (*b*) Circuit symbols.

reversed. In this complementary structure *holes* travel from the emitter towards the collector.

The reader will be interested to learn how this ideal structure is approximately achieved (Fig. 5.3) by using epitaxial growth, oxide masking and diffusion techniques which were described briefly in Chapter 2. Initially an n-type epilayer is grown over an n^+-substrate with an oxide layer subsequently grown over the epilayer. Thousands of transistors will be made together on this single slice, so that initially many thousands of windows are simultaneously etched in the oxide to define the base areas and permit a p-type diffusion, through the oxide mask, to form the base regions of all the transistors in a batch. After this diffusion, the windows that were etched in the oxide are re-covered with new oxide and fresh, but smaller, areas are etched out to define the emitter regions. An n-type diffusion then enters these smaller areas, which have already had the p-base diffusion, and overdopes with n-type impurities. The emitter usually consists of a set of thin stripes, many micrometres long and a few micrometres wide. We shall see later that it pays to have a long periphery to the emitter and very little material directly underneath. To help the over-doping and control of the two diffusion processes, one will usually chose an n-type impurity that has a high diffusion coefficient compared to the p-type impurity. The base will then remain relatively fixed during the second diffusion process. *Emitter-push* is an effect in which the impurities from the second diffusion push the impurities from the first diffusion forward, but modern techniques with slightly different processing and the correct choice of impurities have prevented this effect and base widths down to $\frac{1}{10}$ micrometre can be made if required.

Oxide masking is used once more to define the areas on to which metal is then evaporated to make the contacts to the appropriate regions. The metalization joins up the separate emitters and covers the base regions to form sets of interleaving fingers (alternately base and emitter). A cross-section through such a structure is shown in Fig. 5.3, though other geometries can be used.

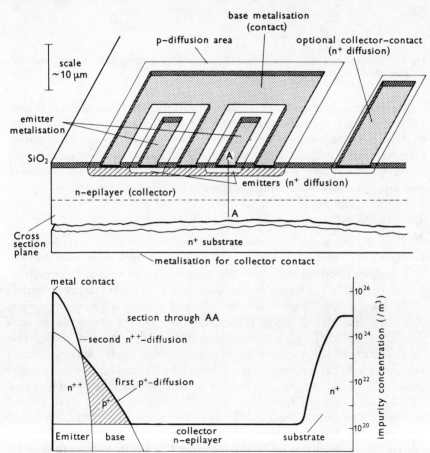

Fig. 5.3 Practical construction of n-p-n transistor (schematic). The contacts to the external leads will be formed by gold wires bonded to the contact pads and to the package leads. Note that hundreds of transistors will be fabricated on a single slice of silicon. The devices shown have only two emitter stripes but one can have over a hundred emitters joined together.

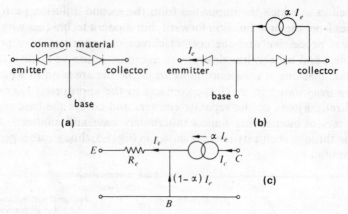

Fig. 5.4   Simple equivalents for a bipolar transistor. (*a*) Two diodes with common base material. (*b*) Ebers-Moll model. (*c*) Low-frequency equivalent circuit in active mode, with many simplifications.

The collector contact is frequently the substrate but sometimes it is convenient to have the collector contact on the top surface along with the contacts to the base and emitter. If this surface–collector contact is required, extra areas are diffused when the emitter diffusion is being carried out. These give relatively low-resistance contacts to the substrate, which in turn is a low-resistance contact for the whole collector area.

The physical action

When the transistor is operating as an amplifier it can, to some extent, be regarded as two p-n junctions placed back to back (Fig. 5.4*a*), but the two diodes are connected by a thin base region of special construction. Indeed the action of two p-n diodes, combined with the transport properties of this base region, forms the basis of a well-known large signal analysis called the Ebers-Moll model (Fig. 5.4*b*). We shall, however, leave such detailed modelling for further reading as required.[5.1] We shall concentrate on the action of the transistor in its normal active mode, where the base–collector diode has a reverse bias so that negligible current flows as a direct result of this bias. Any current flowing in the collector-base diode is caused by electrons, injected from the emitter–base diode, which have *crossed* the base. The emitter–base diode behaves like any asymmetrically doped p-n junction. The high level of impurity in the n-type emitter forces the current, on forward bias, to be carried mainly by electrons from the emitter into the p-type base. The emitter-base current, I_e, will then obey the diode law which is approximately

$$I_e \sim I_0 \exp (eV_{be}/kT) \qquad\qquad 5.2$$

V_{be} is the forward-bias voltage applied across the base-emitter junction. The impedance between the base and emitter terminals is then given by $R_e \sim 26/I_e$(mA) ohms, just like the differential impedance between any ideal p-n junction terminals.

The behaviour of the electrons emitted from the n-region, once they have arrived in the base, forms one of the most important features of the transistor. In the idealized transistor the base has an impurity content varying as $N_{ao} \exp(-\beta x)$, so that the position of the conduction or valence bands is varying with respect to the Fermi-level. This gives rise to an electric field, built into the material by the graded impurity, which prevents the majority carriers (holes) from diffusing to the lower concentration. Equally the field must force any *electrons* entering from the emitter to be driven rapidly across the base towards the collector. Moreover, so long as there is any reverse bias between the base and collector, those electrons which cross the base will be forced into the collector. Hence the base–collector current is determined only by the rate at which electrons are emitted into the base and not by the collector voltage. Thus when an emitter current I_e flows there is a proportion α (close to unity) of this current flowing in the form of electrons which reach the collector *independently* of the magnitude of the collector-base voltage on reverse bias. Hence the current generator shown in Fig. 5.4b,c— with (c) giving one of the simplest but useful 'equivalent' circuits for the transistor in its active mode of operation.

The physical action of the transistor can also be emphasized by the use of energy-band diagrams showing the Fermi levels in the different regions during active operation (Fig. 5.5b). In this case the base is positive with respect to the emitter; the collector is usually positive with respect to the

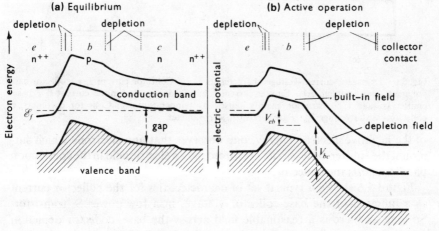

Fig. 5.5 Band diagrams (n-p-n transistor).

base (reverse bias) by a more substantial amount than the base is with respect to the emitter. The applied potential separates the Fermi levels in the contact regions. Remembering that a positive potential is in the direction

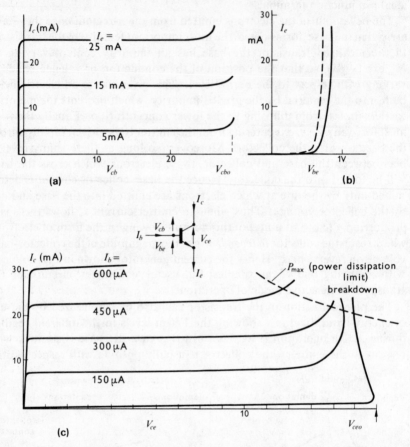

Fig. 5.6 Transistor current-voltage characteristics. (*a*) Collector current I_c v. collector-base voltage V_{cb} (see Problem 5.1). Emitter current I_e as parameter. (*b*) Emitter current I_e v. base-emitter voltage V_{be}. Variation mainly caused by temperature. (*c*) Collector current I_c v. collector-emitter voltage V_{ce}. Base current I_b as parameter.

of the negative electron energy, one observes that any electrons which surmount the emitter-base barrier are forced, by an electric field in the conduction band, towards the collector.

Figure 5.6 shows a typical set of characteristics for the collector current as a function of the base-collector voltage, in a low-power Si transistor. So long as there is a reasonable field across the base–collector depletion region then the collector current is virtually the same as the emitter current (even for small positive-bias voltages on the base–collector junction). The emitter current versus the emitter–base voltage (Fig. 5.6b) has the characteristics of any p-n junction. Current continuity tells one that the current flowing in the base terminal is only $(1 - \alpha)I_e$, a small value. With these factors in mind, the characteristics of the emitter–collector voltage versus collector current, as in Fig. 5.6c can be derived from Fig. 5.6a,b.

The sharply rising current shown in the characteristics of Fig. 5.6c at a voltage V_{ceo} is caused by avalanche breakdown; this process is discussed in Chapter 8. The same process causes the increase in current shown in Fig. 5.6a at voltages larger than V_{cbo}. The fact that V_{cbo} is greater than V_{ceo} is considered in Problem 8.3. At breakdown in the latter example, α has to become effectively very large by charge multiplication, while in the former example it is $\beta \simeq 1/(1 - \alpha)$ that becomes large by α increasing a little. Breakdown also is shown by the emitter–base junction when it is reversed in bias—this breakdown potential V_{beo} is usually small (around a few volts) on account of the high doping concentration which, for small reverse voltages, gives a high electric field in the emitter-base depletion region.

The α factor

In normal operation the α parameter is very close to unity (0·9 to 0·999) and is the product of three principal factors: the emitter efficiency α_e, the base transport factor α_b, and the collector efficiency α_c. The emitter efficiency at a current I_e is defined by $\alpha_e = I_{en}/I_e$ where I_{en} is the current carried by the electrons from the emitter, the remaining current I_{eh} being carried by holes from the base to the emitter. However, it has been seen that in an asymmetrically doped junction the current is primarily carried by the charge carriers from the highest-doped region, so that I_{eh}/I_e is related to the doping ratio N_{ao}/N_{de} appropriate to the base emitter junction (Fig. 5.2a). This ratio is usually small in a well-designed transistor. Hence $\alpha_e = [1 - (I_{eh}/I_e)]$ is virtually unity for the ideal transistor considered here.

The base transport factor, α_b, gives the fraction of current reaching the collector once it has entered the base region from the emitter. An estimate is made here through consideration of the minority carrier charge that is stored in the base. This charge-storage model is capable of considerable development for large signal analysis of transistors,[5.2,5.4] though we shall present only the elements here, with several complicating factors neglected. Initially the recombination of the minority carriers is neglected in the discussion. It is then supposed that the electrons give a principal current I_{cc} flowing from the emitter to the collector. In the base region, these electrons become minority carriers and are driven, in this ideal model, by the built-in field. The time taken to cross the base is τ_b. Thus there is stored in the base a minority-carrier charge $Q_s = I_{cc}\tau_b$. A small recombination rate can now be considered and if the recombination time is τ_r, a current Q_s/τ_r of holes enters at the base contact which can supply holes. However the same number of electrons are required if the charge Q_s is to be maintained. Part of these ($I_{er} = K_1 Q_s/\tau_r$) can be supplied by an increase in the emitter current, and the other part ($I_{cr} = K_2 Q_s/\tau_r$ with $K_1 + K_2 = 1$) can come from a decrease in the collector current. Thus the modified currents, allowing for recombination, now become

$$I_e = I_{cc} + I_{er} = (Q_s/\tau_b) + K_1(Q_s/\tau_r);$$

$$I_c = I_{cc} - I_{cr} = (Q_s/\tau_b) - K_2(Q_s/\tau_r) \quad 5.3$$

so that

$$\alpha_b = I_c/I_e = (\tau_r - K_2\tau_b)/(\tau_r + K_1\tau_b) \approx \tau_r/(\tau_r + \tau_b) \quad 5.4$$

where $\tau_b \ll \tau_r$ has been assumed. This gives an estimate of the base transport factor in terms of the base transit time and recombination rate.

A rough estimate of the base transit time can be made from finding the magnitude of the built-in field E in the base. Assume an impurity profile of $N_{ao} \exp(-\beta x)$; then from eqn 3.30b the valence band potential changes by $(kT/e) \log_e M$ over the base width W where $M = \exp(\beta W)$. The conduction band changes by the same potential, so that there is a field $E = (kT/We) \log_e M$ forcing the electron to cross the base in a time

$$\tau_b = eW^2/(\mu_n kT \log_e M)$$

where μ_n is the mobility. More accurate expressions will be obtained later, when it will be seen that diffusion of the carriers across the base helps, and allows a transistor to work even in the absence of any impurity gradient in the base. Calculations of a few orders of magnitude by the reader will convince him that with $W \sim 10^{-6}$ m, $M \sim 10^3$ then $\tau_b \sim 10^{-10}$ s. Base transit times in the picosecond range are now possible.

The recombination does not take place uniformly throughout the base. For example, in the regions of high doping, especially if the impurity density is high enough to cause dislocation of the lattice, the recombination time will be shorter than in regions of lower doping. Thus the value of τ_r in eqn 5.4 must be an effective value. We shall see later that, for a good switching transistor, τ_r must be kept short, though τ_b should be even shorter. Typically τ_r can have values of $10^{-7} \sim 10^{-9}$ s. The recombination rate and base transit time are substantially constant over a wide range of values for I_c. At very low current densities, recombination assumes a greater importance and one has to account for this by saying that τ_r decreases. At high current densities, one finds that the minority carriers tend to neutralize the majority carriers in the base, so that the built-in field created by the majority carriers decreases. This increases the base transit time τ_b. Thus one finds that at the extremes of current density the α parameter decreases, though there is a wide range over which it is substantially constant.

Finally the collector efficiency α_c is inserted to remind one that the electrons have actually to be collected by the collector contact. With a good construction, the emitted electrons have nowhere else to travel other than to the collector, so α_c is very close to unity. The parameter can, however, exceed unity if the fields in the depletion region are strong enough to cause avalanche multiplication (see Problem 8.3 and Chapter 8). In these cases $\alpha_c = M\alpha_{co}$, with M being the multiplication factor and α_{co} the low field value. The total value for α is then $\alpha = \alpha_e \cdot \alpha_b \cdot \alpha_c$ and, except at very low or very high currents (or breakdown), one takes this value to be virtually constant.

5.3 Low-frequency equivalent circuits

In Fig. 5.4c is shown one circuit which describes the behaviour of the transistor in a simple but not too inaccurate a manner. This circuit has been redrawn in Fig. 5.7a to show how it corresponds to the hybrid circuit for the case when the base terminal is common to both the input and the output (the *common-base* configuration). It may be seen that

$$h_{ib} = R_e \sim (26/I_e \text{ mA}); \quad -h_{fb} = \alpha \sim (0 \cdot 9 \text{ to } 0 \cdot 999)$$

Although we have indicated an output resistance $R_0 = 1/h_{0b}$, both h_{0b} and the feed-back parameter h_{rb} are usually negligible. (It will be seen later that $h_{0b} \sim 10^{-3}(I_e/V_{cb})$ with $h_{rb} \sim 10^{-3}$). The characteristics that are appropriate for this configuration are those of Fig. 5.6a,b. From these it can be seen

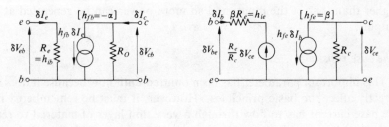

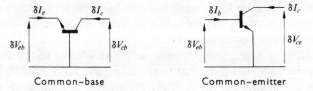

Common–base   Common–emitter

Fig. 5.7 Hybrid-parameter equivalent circuits for transistors (at very low frequencies).

that another way of saying that h_{0b} is negligible is to say that the collector current is virtually constant with changes in the collector–base voltage. In mathematical terms, one says that $h_{0b} = (\partial I_c/\partial V_{cb})_{I_e} \simeq 0$. Similarly h_{rb} can be seen to be negligible because the input characteristics $I_e - V_{be}$ are virtually constant at all collector potentials $[h_{rb} = (\partial V_{eb}/\partial V_c)_{I_e} \sim 0]$.

The parameters h_{fb}, h_{ib} etc. have a suffix b to indicate the common-base configuration as distinct from the arrangement where the emitter is common to both the input and output circuits. A set of hybrid parameters is shown for this latter, more commonly used, configuration in Fig. 5.7b. The appropriate parameters are given by

$$\left. \begin{array}{ll} h_{fe} = \alpha/(1 - \alpha) = \beta & h_{ie} = \beta R_e \sim \beta \,.\, [26/I_e(\text{mA})] \\ h_{re} = R_e/R_c & \text{and} \quad h_{0e} = 1/R_c = \beta h_{0b} \end{array} \right\} \quad 5.5$$

These are found by applying the principles of section 5.1, and it is left to the reader (Problem 5.2) to solve this piece of circuit analysis. It should be noted that one has assumed $R_c \gg \beta R_e \gg R_e$ with $h_{rb} = 0$. The terminology of h_{fe} and β is interchangeable and both are used in the literature. It may also be noted that the output impedance is a factor β lower in the common-base mode than in the common-emitter mode. This can also be seen from the fact that in Fig. 5.6c there is a slight slope on the collector current for a constant base current (see also Problem 5.9). The feedback factor h_{re} is similarly more significant, but even so is often neglected in practical cases.

The amplification properties of the common-emitter circuit may appear more obvious than those of the common-base. Both however can act as amplifiers. In the common-base configuration, although the current is the same in both the input and the output circuit, the output impedance is much higher than that of the input, and so more power can be generated at the output.

The base resistance

This important parameter has been omitted until now because it does not directly affect the basic principles. However, it must be remembered that the base current has to flow through a very thin layer of material to reach the active region where it supplies the required recombination current. There is thus a resistance R_b between the terminal contact for the base and the active region between the emitter and base. This can be significant (~ 100 ohms) and has particular effects at high emitter currents by dropping the base–emitter voltage. Indeed, because the resistance to the centre of the emitter stripe is higher than that to the edge, one finds that the current tends to flow to the edges of the emitter. Non-uniform heating also affects the emitter current distribution. To minimize these effects the emitter is kept with a narrow width and large perimeter. This is particularly important for high-frequency transistors working at high current densities. As far as the low-frequency equivalent circuit in Fig. 5.7b is concerned, one merely has to write

$$h_{ie} = \beta R_e + R_b \qquad 5.6$$

5.4 A practical amplifier

To give the reader a better understanding of the operation of a transistor, a short account is now given of the design of a simple voltage amplifier that could, for example, increase the sensitivity of an oscilloscope if used as a preamplifier. Sophisticated amplifiers will have several stages of amplification and each stage may have more than one transistor.[5.13] Many practical amplifiers would now in fact use Integrated Circuits rather than discrete components so that all the user would require would be a power supply. Nonetheless, looking at the design of a single-transistor amplifier will help the reader understand many practical points in the use of transistors.

Stabilizing the operating point[5.14]

To make a transistor work in its active region it has to be supplied with current and voltage. One also requires the changes of output voltage to be kept in a constant relation to the changes at the input; these amplifying characteristics should not change with time or temperature, or with changes of transistor (of the same type). It is found that if these requirements are met the transistor must then operate with a well-controlled mean value $(\bar{I}_c, \bar{V}_{ce})$ as its 'operating point'. The analysis of a simple circuit (Fig. 5.8) which will achieve a useful result will be instructive.

Any design is usually achieved by successive considerations, so let us first consider how to fix the mean collector current and voltage values at $(\bar{I}_c, \bar{V}_{ce})$ as above. In principle the base current $\bar{I}_c/\beta$ could be defined and fixed. However β varies among devices, even of the same type, by a factor of 3 to 1; the base current cannot therefore be closely defined. Temperature affects the base-emitter voltage V_{be} for a given current I_e, so that fixing V_{be} is not a good technique. There is also another source of uncertain magnitude, not yet discussed, that adds to the base current—the reverse saturation current for the collector–base junction.

Like any p-n junction on reverse bias, the collector–base has a small reverse current which is a hole current for the n-p-n transistor under discussion. This current flows into the base and cannot be distinguished from the hole current that flows from the base terminals. Consequently, if the terminal base current is I_b then the effective base current is $I_b + I_{cbo}$, where

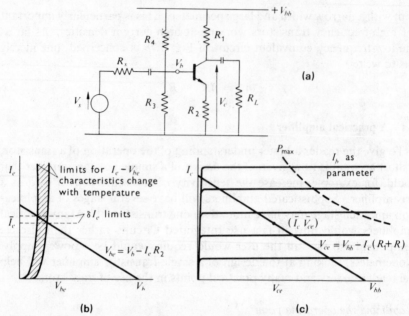

Fig. 5.8 A simple amplifier. (*a*) Circuit layout. V_s is source; R_L is load. (*b*) Stabilization of I_e. (*c*) Load line indicating a choice of operating point.

I_{cbo} is the reverse saturation current. The collector current is then $\beta(I_b + I_{cbo})$. Fortunately for circuit designers, although I_{cbo} is uncertain it is often negligible in modern silicon transistors. Typical values are less than 10^{-8} A, though this value roughly increases by a factor of 2 for every 8 °C rise in temperature. Thus at elevated temperatures βI_{cbo} can be significant. In power transistors, especially if they are made of germanium where I_{cbo} is more significant, the emitter and collector current can show considerable thermal instability if not properly controlled.

We are thus left with fixing the emitter current as a good way of stabilizing the operating point (I_c closely follows I_e). It should be noted that in Fig. 5.8*a* $\bar{V}_{be} = V_b - \bar{I}_e R_2$. This is a line that can be drawn on the emitter–base characteristics (Fig. 5.8*b*) and it can be seen that changes in this characteristic by a few hundred millivolts in V_{be} will result in very small changes in $\bar{I}_e$, given that $\bar{V}_b \sim 3$ V. The potential of $\bar{V}_b$ can be fixed by the potential divider $R_3 : R_4$, say at 3 V, so that $\bar{I}_e \simeq 2{\cdot}4/R_2$ to within 10 per cent for a Si device with $V_{be} \sim 0{\cdot}6$ V. One must note that if the chain $R_3 : R_4$ is to fix V_b† then the chain current $I_{ch} = V_{bb}/(R_3 + R_4)$ should be several times the base current I_e/β [$I_{ch} \sim (10/\beta)I_e$ will suffice]. There are other ways of fixing the operating current which do not need a chain, but this is left as a Problem (5.3).

† $V_b \simeq V_{bb} R_3/(R_3 + R_4)$.

Choice of load and current

The actual value to be chosen for $\bar{I}_e$ now needs consideration. The collector–emitter voltage $\bar{V}_{ce}$ is given by noting that $\bar{I}_c \sim \bar{I}_e$ and

$$\bar{V}_{ce} = V_{bb} - \bar{I}_c(R_1 + R_2)$$

This is a *load* line that can be drawn on the characteristics of Fig. 5.6c, as in Fig. 5.8c. This line cuts the characteristic at one point determined by the base current. One can see that as the base current changes so the voltage across the device will change. It will be advantageous to bias the device with approximately $\bar{V}_{ce} = \frac{1}{2}V_{bb}$ so that the greatest change of voltage can be made in either direction. The operating point must be chosen so that the maximum power limits are not exceeded, and in general this implies that

$$V_{bb}{}^2/4(R_1 + R_2) < P_{\max}$$

The value of $P_{\max}$ is given in the manufacturers data. Again it is advisable to have V_{bb} lower than V_{beo}, the breakdown value. Within these limits there is a wide range of acceptable conditions. The value of $\bar{I}_c$ will then probably be chosen by the load required of the amplifier. If the amplifier has to deliver power to a load of resistance R_L, it will be found useful to keep the value of R_1 considerably less than R_L. It will be noted from Fig. 5.8 that R_L is connected by a capacitor; this is sufficiently large to have a small impedance at any frequency of interest. The signal source is similarly connected. Thus with R_1 chosen, V_{bb} fixes I_c and I_e, the stabilization of I_e fixes V_b and R_2 to some extent, and V_b, V_{bb} and I_c/β fix the chain values R_3 and R_4. Knowledge of how the amplifier actually works will help to narrow the choice further.

Operation of the amplifier

As far as *changes* in the voltage are concerned, the power supply remains at a constant potential and so appears to be a perfect a.c. short circuit. This allows the circuit to be redrawn (Fig. 5.9) with the equivalent circuit for the transistor ($h_{re} = h_{oe} = 0$ assumed). The resistance R_3 and R_4 now appear as a parallel

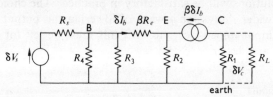

Fig. 5.9 Equivalent circuit for transistor in amplifier. In transistor circuit, base resistance R_b has been ignored. In analysis, R_3 and R_4 assumed very large; R_L not connected in first instance to show gain $\delta V_c/\delta V_s$ can approximate to $-R_1/R_2$.

combination, but even so these are usually high enough in impedance to be ignored (as they are here). Straightforward analysis yields

$$\delta V_c = -R_1\beta\,\delta i_b; \quad \delta V_s = \delta i_b(R_s + \beta R_e + R_2) + \beta\,\delta i_b R_2 \qquad 5.7$$

Hence, provided the source impedance is low enough and β high enough,

$$\delta V_c/\delta V_s = -R_1/(R_2 + R_e) \sim -R_1/R_2 \qquad 5.8$$

This is the voltage gain and can be seen to be independent of the precise value of β and other device parameters. The negative sign shows the inversion of the output: a rise in voltage at the input results in a fall at the output. A more detailed expression for the gain shows that the output voltage falls by a factor of 2 if the source impedance becomes comparable with $\beta(R_e + R_2)$. This value is the *input impedance* of the amplifier and it can also be calculated from the ratio of the input voltage divided by the input current, at the input *terminals* of the amplifier.

With the load R_L connected across the output, the voltage falls by a factor $R_L/(R_1 + R_L)$. This fall is a factor of 2 if $R_L = R_1$ and R_1 is the *output impedance* of the amplifier. In general one tries to operate sections of amplifiers with input impedances that are lower than the source impedances and similarly with output impedances that are lower than the load impedances. This protects the amplifier characteristics from drastically changing when the source or load is slightly changed.

The voltage gain of this amplifier could be greatly enhanced by placing a by-pass capacitor C_2 across R_2. The steady-state operating point would not be altered. However, for changes of current and voltage one would place R_2 at zero. With $R_s = 0$ and $R_2 = 0$ one finds that a first-order estimate is given by the gain $-R_1/R_e = -(R_1\bar{I}_e)/(kT/e)$. The base resistance (that has been neglected in the equivalent circuit) usually means that a lower value than this is found in practice. One notes then that the gain will be sensitive to the temperature and other transistor parameters. High-frequency effects will reduce the gain in most single-stage amplifiers, though these are not discussed in this simple analysis. The reader should see then that there is no unique design procedure. A number of factors must be considered and a number of solutions will be satisfactory in practice. The choice of circuit leads to a far wider discussion—ways of lowering the output impedance or raising the input impedance, etc. All this must be left for reading on circuit analysis and design[5.13,5.15].

5.5 Base transport

We now turn to a more advanced topic in transistor theory. The purpose of this section is to show that diffusion can play an important role in the base transport and that graded impurity profiles are not essential in the base. The detailed theory will also allow one to make a more accurate evaluation of the h-parameters, h_{ob} and h_{rb}, that were previously neglected.

If recombination is initially neglected, then in the n-p-n transistor no hole current will flow in the base and $J_{cp} = 0$, so that if $p = N_a$, eqn 3.13b will imply that there is a built-in field

$$E = (D_p/\mu_p) \frac{1}{N_a} \frac{\partial N_a}{\partial x} \qquad 5.9$$

The electron convection current density J_n is constant with distance through this base (no recombination) so that eqn 5.9, with the constancy of J_n, allows one to integrate eqn 3.13a as a first-order differential equation in $n(x)$:

$$n(x) = n(W)[N_a(W)/N_a(x)] + [J_n/eD_nN_a(x)] \int_x^W N_a \, dx \qquad 5.10$$

Note that the integration has been done from the collector edge of the base $(x = W)$ backwards. At the collector edge, the electric fields rise very rapidly and quickly pull any electrons across the base-collector depletion region. In fact the fields are so high that the electrons often move close to their limiting velocity $v_s (\sim 10^5$ m/s see p. 179). This then gives $n(W) \sim J_n/ev_s$, though many authors put this value $n(W)$ at zero as an approximation. We have retained a finite limit on this edge density because we believe that it is more realistic for really high-performance transistors operating in the microwave frequency range. Using $n(W) = J_n/ev_s$ and putting $N_a(x) = N_{ao} \exp(-\eta\chi)$ one can evaluate $n(x)$ from eqn 5.10 (see Fig. 5.10). In particular

$$n(0) = (J_n/Fev_s)[1 - (1 - F) \exp(-\eta W)] \qquad 5.11$$

where $F = \eta D/v_s$. The factor F cannot exceed unity because as η increases so the built-in fields approach the value at which the limiting velocity is reached. The mobility relation of field and velocity then breaks down and so does the Einstein relation used in eqns 5.9 and 5.10. One may note that if $F = 1$ then $n(x) = n(W)$ for all x—the uniform distribution of carriers that was considered in the idealized model.

We shall assume that the value of $n(0)$ is controlled by the usual Boltzmann relation for minority carriers at the edge of a p-n junction:

$$n(0) = (n_i^2/N_{ao})[\exp eV_{be}/kT - 1] \qquad 5.12$$

This, with eqn 5.11, gives the emitter current density as a function of base width and emitter–base voltage.

The base transit time can be evaluated directly. In eqn 5.3 it was asserted that $\tau_b = Q_s/I_c$, where Q_s was the charge of electrons stored in the base.

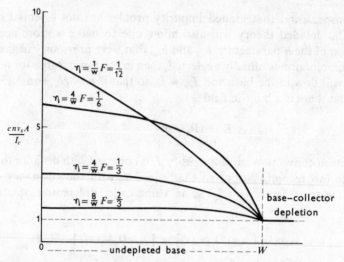

Fig. 5.10 Schematic charge distributions in graded base transistor (allowance has been made for the limiting velocity v_s of the charge carriers: $F = \eta D/v_s$ with base impurity

$$N_s = N_{as} \exp(-\eta x).$$

This can be seen more clearly perhaps by noting that the average velocity, including both diffusion and field, is given by the average rate of particles crossing a plane: thus $\bar{v}_n = (J_n/en)$. Hence

$$\tau_b = \int_0^W (1/\bar{v}_n)\,dx = (1/J_n)\int_0^W en\,dx = Q_s/I_c \qquad 5.13$$

($I_c = J_n$. Area, $I_c = I_e$ in absence of recombination). For the example corresponding to the conditions for eqn 5.11,

$$\tau_b = (W/Fv_s) + [(F - 1)(1 - \exp -\eta W)/(F\eta v_s)] \qquad 5.14$$

As $F \to 1$ so $\tau_b \to W/v_s$, as would be expected. Note also if $\eta = 0$ then the base impurity is uniform; allowing $\eta \to 0$ gives $\tau_b \to W^2/2D_n$, and this is the example dealt with extensively in many textbooks but omitted here in favour of the model with a strong built-in field.

The basic reason for the variation of the emitter current with the collector potential is that changes in the collector–base potential alter the depth of depletion in the base and so change the undepleted base width, W. A rough estimate can be made of $\partial W/\partial V_{cb}$† by noting that for a one-sided p-n junction the depletion varies as $d_0(V_{cb}/V_0)^{1/2}$, where the depletion depth is d_0 at the voltage V_0 (see eqn 4.5a). The depletion d' on the highly doped p-side is shorter by a factor of the doping ratios, so that $d' \simeq d_0[N_{dc}/N_{dw}](V_{cb}/V_0)^{1/2}$.

† W is in fact assumed to be only a function of V_{cb}, so that in strict mathematical terms this expression is dW/dV_{cb}.

The change of base width with collector voltage is thus given by $\partial W/\partial V_{cb} = -\partial d'/\partial V_{cb} = K(d_0/V_{cb})$ where $K = \frac{1}{2}(N_{dc}/N_{aw})(V_{cb}/V_0)^{1/2}$ a parameter that can be made to be of order 10^{-1} by choice of doping ratios. The modulation of base width with changes of collector voltage was first recognized by J. M. Early and is known as the Early effect.[5.3] It is far more serious in older types of transistor where the collector doping was much higher than the base doping. The Early effect is largely responsible for the voltage feedback ratio h_{rb} and the output impedance h_{ob} that we previously neglected. To show the relationships, one uses standard differential calculus as follows.

First note that $h_{rb} = (\partial V_{be}/\partial V_{cb})_{I_e}$ and this can be rearranged to

$$h_{rb} = (\partial I_e/\partial V_{cb})_{V_b}(\partial V_b/\partial I_e)_{V_c} = R_e(\partial I_e/\partial V_{cb})_{V_{be}} = R_e(\partial I_e/\partial W)_{V_{be}}(\partial W/\partial V_{cb})$$

However $I_e = J_n$. Area so that eqns 5.11 and 5.12 allow evaluation of $(\partial I_e/\partial W)_{V_{be}}$ to give

$$h_{rb} = K(I_e R_e/V_{cb})(1 - F)[\eta d_0/\exp \eta W)] \qquad 5.15$$

where K is given above by the evaluation of $\partial W/\partial V_{cb}$. For a good base structure $(1 - F)/\exp(\eta W)$ is so small that h_{rb} often may be neglected, taking values around 10^{-3}.

The parameter h_{ob} is similarly determined from $\partial W/\partial V_{cb}$ through

$$h_{ob} = -(\partial I_c/\partial V_{cb})_{I_e} = -I_e(\partial \alpha/\partial V_{cb}) = -I_e(\partial \alpha/\partial W)(\partial W/\partial V_{cb})$$

It is now assumed that recombination is a small perturbation which hardly affects the distribution of the charge, so that α is determined by the base transport time τ_b as in eqn 5.3b; then $\partial \alpha/\partial W \simeq -(1/\tau_r)(\partial \tau_b/\partial W)$ and at large enough values of (ηW) this gives, from eqn 5.14,

$$h_{ob} \sim (I_e/V_{cb})(K d_0/\tau_r F v_s) \qquad 5.16$$

Recombination times around 10^{-7} s, with depletion regions around 10^{-5} m, will give values of h_{ob} around $10^{-3}(I_e/V_{cb})$, an often negligible value. However, with transistors designed to operate in the microwave range of frequencies, it is difficult to keep a good base profile over the very thin base, and moreover τ_r is usually in the nonosecond range. In such devices the Early effect is significant and cannot be overlooked as in the elementary design work presented in this book.[5.4,5.6]

5.6 The high-frequency performance of the transistor

To round off the discussion on the transistor, the charge control equations given by eqn 5.3 are developed to show how time-dependence affects the performance. The point to remember is that the base current from the base contact, in the *ideal* transistor, supplies only the majority carriers required to recombine with the minority carrier charge Q_s stored in the base. In the n-p-n transistor, a change dQ_s in the charge of electrons stored in the base cannot be supplied by the base current, which will be of holes, but must be given by changes in the emitter and collector currents. An extra electron current dQ_s/dt is required to supply a change dQ_s in a time dt. Thus eqn 5.3 is modified to:

$$I_e = (Q_s/\tau_b) + K_1(Q_s/\tau_r) + k_1(dQ_s/dt) \qquad\qquad 5.17a$$

$$I_c = (Q_s/\tau_b) - K_2(Q_s/\tau_r) - k_2(dQ_s/dt) \qquad (k_1 + k_2 = 1) \qquad 5.17b$$

The relative proportions, $k_1:k_2$, of the current dQ_s/dt which are supplied from I_e and I_c require detailed dynamics for their calculation, and indeed would depend in general on the frequency of operation. Fortunately, like the proportions $K_1:K_2$, these detailed values will not need to be known and only the fact that $k_1 + k_2 = 1$ will be required. It must also be observed that Q_s determines I_c and not vice-versa, so that there is never any question of a complementary function growing exponentially with time as a solution to eqn 5.17b. Inserting a frequency dependence $\exp j\omega t$ and using $\tau_r \gg \tau_b$,

$$\alpha(\omega) = I_c/I_e \simeq \alpha_0(1 - jk_2\omega\tau_b)/(1 + jk_1\omega\tau_b) \qquad\qquad 5.18$$

with α_0, the d.c. value given in eqn 5.4, multiplied by the emitter and collector efficiencies which are assumed to be frequency independent.

The equivalent circuit of Fig. 5.4c must now be reconsidered. The charge stored in the base, Q_s, can be determined from eqns 5.10 and 5.12, and so from eqn 5.12 is a function of V_{be}. Typically in the ideal model, where the reverse current from the collector is ignored, $Q_s = Q_{so}[\exp(V_{be}/kT) - 1]$. Consequently, as in the p-n diode, a diffusion capacitance C_d must be added. In the transistor, this capacitance must appear across the emitter resistance R_e of Fig. 5.4c. To establish the correct time constant it is required that $C_dR_e = k_1\tau_b$; this will then take care of the charge storage term in eqn 5.17a. The charge-storage term in eqn 5.17b is accounted for in the complex value of α given in eqn 5.18. Now, as well as the direct transport of charge, there must be the current flow arising from the depletion capacitances C_e (from base to emitter) and C_c (from base to collector). Figure 5.11a shows these changes with the transistor drawn as for a common-emitter configuration. It will now be shown that the h-parameters for the common-emitter configuration can typically take on a fairly simple form.

Note that $1 - \alpha = (1 - \alpha_0 + j\omega\tau_b)/(1 + j\omega C_dR_e)$. This will ease the analysis of the circuit. Then define a transitional frequency ω_t such that $1/\omega_t = [\tau_b + C_eR_e + C_cR_e]$. We shall find this characteristic frequency occurring time and again and especially in the combination of ω_t/β_0 where $\beta_0 = \alpha_0/(1 - \alpha_0) \simeq 1/(1 - \alpha_0)$. The whole simplification of the analysis

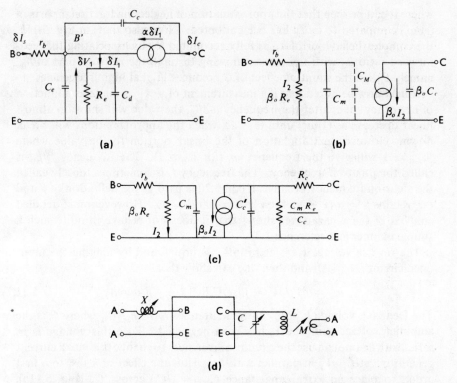

Fig. 5.11 High-frequency effects in transistors. (a) Development of the common base. (b) Common-emitter circuit at intermediate frequencies. The voltage gain G of the amplifier leads to a Miller capacitance $GC_c = C_M$ across C_m. (c) High-frequency circuit with low gains and omitting the reactances caused by the package around the transistor and the leads to the device. (d) An oscillator circuit: C controls the frequency, M and X adjust the transfer of power to the base-emitter circuit from the collector.

arises on the assumption that the useful operating frequency for the transistor lies below or around the frequency $f_t = \omega_t/2\pi$, so that fractional parts of (ω/ω_t) can be ignored compared to unity. The h-parameters are found in the usual manner. Let us begin by shorting the collector so that $\delta V_{ce} = 0$. It can then be shown that $\delta V_1/\delta I_b = \beta_0 R_e/[1 + j(\beta_0\omega/\omega_t)]$ so that

$$h_{ie} = r_b + [\beta_0 R_e/(1 + j\omega\beta_0 R_e C_m)] \qquad 5.19$$

where $C_m = (1/\omega_t R_e)$. This then is equivalent to the resistance–capacitance combination shown at the input in Fig. 5.11b. The forward current transfer ratio is

$$h_{fe} = \frac{\delta I_c}{\delta I_b} = \frac{\alpha_0 - jk_1\omega\tau_b - j\omega C_c R_e}{1 - \alpha_0 + j\omega\tau_b + j\omega C_c R_e + j\omega C_e R_e} \simeq \frac{\beta_0}{1 + j(\beta_0\omega/\omega_t)} \qquad 5.20$$

where it can be seen that the approximation of neglecting fractional parts of (ω/ω_t) compared to *unity* has been adopted. Note also that, in Fig. 5.11b, the complex behaviour of h_{fe} is fully accounted for by associating the feed-forward ratio β_0 with the current flowing through the *resistive* part of h_{ie}, namely $\beta_0 R_e$; the shunting effect of C_m reduces $|h_{fe}|$ at high frequencies.

It may now be noted that if a measurement of $|h_{fe}|^2$ is made as a function of frequency then, after the frequency ω_t/β_0, the value will show an almost linear decrease as $(1/\omega^2)$ until $\omega \sim \omega_t$ when the approximations will break down. However, extrapolation of the linear portion to the value where $|h_{fe}| = 1$ will give the frequency ω_t (for $\beta_0 \gg 1$). The frequency f_t/β_0 is called the β-cut-off frequency. The frequency f_t is sometimes loosely called the α cut-off frequency because, if eqn 5.20 is used as the definition of β and one retains $\beta = \alpha/(1 - \alpha)$, then $|\alpha|^2 = \frac{1}{2}$ at $\omega = \omega_t$. However more detailed analysis of the α parameter shows that it does not behave in quite such a simple manner (see reference 1.2).

The reverse voltage transfer ratio/h_{re} is now found by making the input open circuit ($\delta I_b = 0$) and then one can show that

$$h_{re} = \delta V_1/\delta V_c = j\beta_0 \omega C_c R_e/[1 + j(\beta_0 \omega/\omega_t)] \qquad 5.21$$

The feedback voltage $h_{re} \, \delta V_c$ can be written as $j\omega G C_c h_{ie} \, \delta V_1$ where G is the amplifier voltage-gain and r_b has been neglected. Since this voltage is in series with h_{ie} one can use the circuit equivalences to change this into a current generator $j\omega G C_c \, \delta V_1$ in parallel with h_{ie}. Thus the effect of h_{re} is, to a first order, to place an extra capacitance $C_M (= G C_c)$ across C_m (Fig. 5.11b). This is often referred to as the Miller capacitance.

Within the approximations given previously, $\delta V_1 = \delta I_c R_e$; thus

$$h_{oe} = \delta I_c/\delta V_c = (C_c/C_m R_e)j(\beta_0 \omega/\omega_t)/[1 + j(\beta_0 \omega/\omega_t)] \qquad 5.22$$

At fairly low frequencies the output conductance is ignored and a capacitance of approximately $\beta_0 C_c$ is added (Fig. 5.11b), while at very high frequencies the output capacitance is lower but the output conductance approaches $C_c/C_m R_e$ as in Fig. 5.11c. At very high frequencies it may be necessary to include a collector–contact series resistance R_c which, although small, limits the speed at which the collector depletion capacitance can charge and discharge. However, the reader must appreciate that, at such frequencies, the package parasitic capacitances and lead inductances often dominate the considerations for the circuit design.

The highest theoretical frequency at which the transistor can be of any practical use is called f_{max}. This is the highest frequency at which a lossless circuit could be made to oscillate. In Fig. 5.11d such a circuit is shown schematically. In principle L and C are chosen so as to tune out the reactances in the collector–emitter circuit. If the collector–contact resistance R_c is neglected, then $(\beta_0 I_2)^2 (C_m R_e/8C_c)$ is the maximum power that can be transferred from the transistor into a load (assuming the collector-emitter resistance given by the limiting value of eqn 5.22 at large ω). In principle then the mutual coupling M and the reactance X can be chosen so as to arrange for maximum power transfer into the base–emitter circuit. The frequency f_{max} is sufficiently far above ω_t/β_0 for the power to be absorbed only in the base resistance r_b. The reactances again are all tuned out. For oscillation just to occur, one requires the maximum power out to equal the power in. The power in, however, is approximately

$$\tfrac{1}{2} r_b (I_2 \omega_{max} C_m \beta_0 R_e)^2$$

so that
$$f_{max} = \omega_{max}/2\pi = (1/4\pi)(\omega_t/r_b C_c)^{1/2} \qquad 5.23$$

One can see that at this frequency f_{max}, the power gain of the transistor has fallen to unity. At higher frequencies it merely acts like an attenuator!

For very-high-frequency transistors, the value of ω_t must be modified to allow for the transit time τ_c of the carriers across the collector–base depletion region, and also to allow for the series resistance R_c in the collector contact giving and $R_c C_c$ time constant for charging and discharging the collector. Then $1/\omega_t \simeq (\tau_b + \tau_c + R_c C_c + R_e C_c + R_e C_e)$. In principle, it would appear that the lowest value of this overall time constant would occur at really high currents when $R_e(C_e + C_c)$ was negligible. However, at very high current densities the base transit time is increased because the built-in fields become less effective due to neutralization of the majority carriers by the injected minority carriers. Thus there will be an optimum value of I_e for a high f_t. The design compromises for very high-speed transistors are many (Problem 5.8), and are well described in reference 5.6.

The transistor as a switch[5.7,5.16]

Transistors are widely used in high-speed computers because they can act as switches. Variations in the base current are used to switch the impedances between the emitter and collector terminals from a high to a low impedance state. The speed of this switching depends on the circuit configuration, but for the popular common-emitter configuration, where there is current gain, the speed is typically determined by $1/\tau_r$ (τ_r the recombination time) or equally f_t/β_0. One must remember that the state of the transistor can only be changed if the minority-carrier stored charge in the base can be changed. The minority-carrier charge is controlled by the base current, thus from $I_b = I_e - I_c$ and from eqn 5.17

$$I_b = (Q_s/\tau_r) + dQ_s/dt$$

If then there is a step change from I_{b1} to I_{b2}, as shown in Fig. 5.12a with $R_L = R_1$, then

$$Q_s = I_{b1}\tau_r \exp -t/\tau_r + I_{b2}\tau_r(1 - \exp -t/\tau_r) \qquad 5.24$$

Given that τ_b is very much shorter than τ_r, I_c will closely follow Q_s, as in Fig. 5.12c. Waveforms sometimes have 'wiggles' in them because, as stated earlier, τ_r is not actually uniform throughout the base. Consequently the true waveforms cannot be represented by a single time constant. Changing

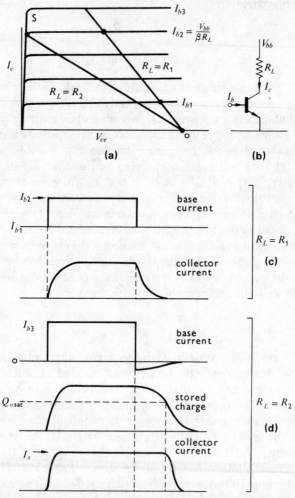

Fig. 5.12 Current and saturation mode switching. (a) Load lines: R_1 for current and R_2 for saturation mode. (b) Simple switching circuit. (c) Current relationships with time for current mode switching. (d) Current and charge relationships for saturation mode switching.

the collector current between two current levels, where the transistor is always in its active state, defines the *current mode* of switching. When combined with the circuit configuration considered in Problem 5.5 this mode gives the fastest switching speeds and must be contrasted with the saturated mode discussed below.

More precise definition of collector voltages, with less dependence on the value of β, is found in the *saturated mode* of switching. In Fig. 5.12a, $R_L = R_2$ and $I_{b3} > V_{bb}/\beta R_L$ so that the transistor has its operating point S at a very low collector voltage. Indeed the load absorbs nearly all the voltage so that one finds the collector–base junction is on *forward* bias with the collector *below* the base potential. The transistor is no longer active but is approximated to by an equivalent circuit which assumes all terminals are shorted together. Typically, one finds in an Si transistor that V_{ce} is about 0·2 volts while V_{be} is still 0·6 to 0·7 volts. This is the *saturated state* (it is emphasized that the saturation current $I_{c\,sat}$ depends on the load: $I_{c\,sat}R_L \simeq V_{battery}$). The load rather than the collector–base depletion region is then absorbing the voltage. In saturation there is more charge stored in the base than is required for normal active mode operation; the collector–base junction, on forward bias, stores charge as well as the emitter-base junction. Thus before the transistor can be switched out of saturation the excess charge must be removed, and this causes a delay between the applied voltage at the base and the change of collector voltage. This delay is shown schematically in Fig. 5.12d. It can be above a microsecond in poor transistors. However, modern switching transistors tend to clamp a Schottky diode across the collector–base junction, in the same polarity (see Fig. 7.3). This type of diode is discussed in Chapter 7; it will be seen that any forward bias current will be diverted through the Schottky barrier rather than the Si transistor. This cuts out the base storage effects in saturation and gives much faster switching.

Let us now consider the other end of the load line in Fig. 5.12a—the point labelled O. Here there is no base current and the base–emitter junction is usually on reverse bias or at least zero bias. In this case both diodes of the transistor structure are in the high-impedance reverse-bias state. An approximate equivalent circuit for the transistor in this state is given by all three terminals isolated. Saturated mode switching leads then to well-defined voltage changes at the collector, typically from zero to close to the battery voltage. Switching times can be in the nanosecond range with Schottky clamped transistors. This good definition of the voltages makes the device ideal for logic circuits in high-speed computers.

There is another mode of switching the transistor, the *avalanche mode*, where high voltages create high electric fields in the base–collector region and cause avalanche breakdown as discussed in Chapter 8. The extra charge generated by the breakdown changes the output impedance from a high value to a very low value in sub-nanosecond times. The change is triggered by a small injection of charge at the base. The transistor has to dissipate a lot of power and only operates at low-duty cycles. This can be useful in special applications not discussed here.

5.7 The monolithic integrated circuit

The integrated circuit (IC) was briefly mentioned in Chapter 1. It is difficult to overestimate both its practical importance and its impact on the use of semiconductors. The variety of available ICs which will perform different functions grows at a rate which only an encyclopaedic mind can follow. This book does not attempt to show this variety of circuit function but merely to indicate how, in principle, some discrete devices can be formed, in a batch process, into a functional circuit built on a single piece of semiconductor material. The importance of the *batch processing* must be emphasized because it is only by producing thousands of identical circuits simultaneously, rather than sequentially, that the benefits of low cost and uniformity are obtained. Each process step must have a very high *yield;* in other words, the proportion of failures introduced by the step must be very low. We shall outline here a few of the processes which have been successfully applied to silicon in order to make ICs.

It will be assumed that a circuit with some specific function has already been chosen and that transistors, diodes, resistors and capacitors will form the basic elements. Other devices may be used and may require different or even simpler technologies than the one outlined; nevertheless some of the basic ideas will be similar. The reader will have already seen (section 5.2) how thousands of transistors can be simultaneously made, in a series of batch processes, on a single slice of silicon. However, as the process is outlined in section 5.2, the silicon n-substrate would electrically connect to all the transistors; what we need now is the electrical isolation of the transistors. Four different techniques for achieving the isolation are illustrated in Fig. 5.13. To avoid too many details, most of the particulars of the construction of the n-p-n transistors are not repeated or illustrated and can be taken to be similar to the construction shown in Fig. 5.3, with the top collector contact. Complementary processes are in principle possible to produce p-n-p transistor circuits. The first method shown (Fig. 5.13a) is that of *diode isolation*. Here a tub of n-type material is surrounded by p-type material which is then connected to the most negative point of the whole circuit. Adjacent tubs are then isolated from each other by the depletion capacitance that forms between the p and n-regions on reverse bias. The buried layer of n^+-material that is shown forms the rear collector terminal like the n^+-substrate in Fig. 5.3. This layer can be made by a selected area diffusion or epitaxial growth before the n-epilayer is grown over the p-substrate. A deep p-diffusion completes the isolation of the n-type islands, and the electrical contacts to the transistors are then all formed on the top

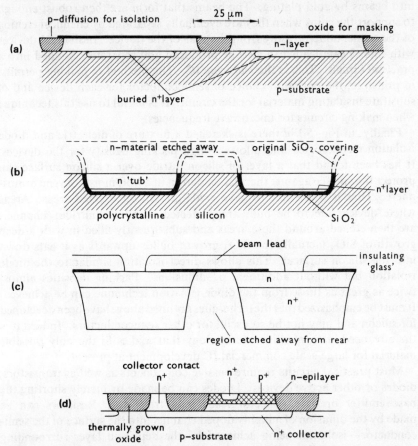

Fig. 5.13 Isolation techniques for integrated circuits. (a) Diode isolation. (b) Dielectric isolation. (c) Air isolation (beam lead). (d) Oxide-isolated epitaxial growth.

surface with the help of photolithography and evaporation. Figure 5.13b shows just one example of the many isolation techniques which use *dielectrics*. In this example, islands of n-type material (perhaps with n^+-epilayer on top) are etched out from a disc of silicon. The whole etched surface is then covered with some insulating glass such as SiO_2. Polycrystalline silicon can then be grown on the back of this glass in an epitaxial reactor (note that the growth is polycrystalline because there is now no ordered lattice for the growth to start and develop upon). The main part of the n-material is then removed, leaving 'tubs' of the single crystalline material embedded in a matrix of polycrystalline material but insulated by the oxide film. In Fig. 5.13c is shown one of the air isolation techniques—*beam lead* technology. The electrical connections are made by evaporation of metallic layers built up into beams by gold plating. The beams that form are then robust enough to support the chips when they are eventually separated by chemical etching of the semiconductor away from the rear of the device. Individual devices with short beam leads can also be produced and separately bonded into a practical circuit. This is the *multichip* IC. The method of making permits, in principle, the optimum choice of semiconductor for each device and of substrate insulating material for the circuit. It is useful to use this technique when making circuits for microwave frequencies.

Finally, in Fig. 5.13d there is sketched a mixture of dielectric and diode isolation technologies that allows a high packing density for the devices. It has been found that a layer of silicon nitride over a silicon surface can protect the surface against thermal oxidation in a hot, moist, oxygen atmosphere. Thus it is possible to have oxide growth on selected areas. Areas where devices are to be built are protected by silicon nitride; channels are then etched around these areas and subsequently filled in with a deep growth of SiO_2 (actually the SiO_2 growth builds upwards as it eats down into the silicon surface). This allows direct isolation similar to the diode isolation but without additional p^+-diffusions. Packing densities almost twice as great as those from the diode-isolation technique can be achieved. It must be emphasized that the techniques outlined above have been developed for silicon and may not be so useful for other semiconductors. Indeed it is the advanced state of silicon technology that makes Si the only possible material for large-scale commercial IC development at present.

Most practical circuits require passive components as well as transistors, diodes or other active devices. Diodes can be made by merely shorting the base–emitter or base–collector leads in a transistor. Resistors can be made by the diffusion of a highly doped channel into the surface of the semiconductor—isolation being achieved by the depletion layer surrounding a p-channel in a n-block for example. Metal films can be used as an alternative. By either method the range of resistance that can be reliably reproduced is limited, either by the tolerances of producing small areas with photolithography or by the large area required if the resistance is very low. In general, the circuit designer can rely on well-maintained ratios for the resistance values, though not on absolute resistance. Capacitances can be produced by the junction capacitance between p and n materials, or again by metal films over oxide layers. As for resistors, the range of values permitted is limited by area and definition, and the circuit designer must know these limitations.

Once the isolation technique has been chosen, the method of electrical interconnection must follow. In the beam lead system the interconnection is an integral part of the isolation process. With the diode isolation the interconnections are usually formed by evaporation of metal (often Al) patterns connecting up the isolated devices. Oxide masking is again used here, and acts as an insulator over the surface of the IC except in the exposed, correct, contact areas. This batch-production technique of connecting devices clearly has many benefits over the individual attention required to a discrete component circuit. The reader may like to try listing these; e.g. leads do not vibrate loose, cost is lower etc.

Finally the circuit is packaged. Gold wires are bonded to the contact areas and to the package leads by a process called thermocompression bonding (TCB).† The device is then embedded in epoxy resin to form a package that is much larger than the chip—a typical size for a chip containing some 100 devices may be 250 micrometres square.

We conclude this section by comparing two circuits in discrete and IC form (Fig. 5.14) to indicate briefly some of the ways in which the IC has altered design philosophy. The first circuit is an example of a logic circuit using saturated mode switching. In these circuits there are two voltages referred to as logical 0 and logical 1. If any of the inputs A, B or C is down at logical 0 (close to zero volts) the design of the base resistor will ensure that one of the transistors T_1 will be saturated. The collector voltage of T_1 (which also corresponds to the base voltage of T_2) is then so low that T_2 is in the off state and so the voltage G is up at a value of V_{bb}. This takes V_0 up to logical 1, which is about a volt below V_{bb} in this case. If however all of A, B and C are up at logical 1 then T_1 will be in an active state and can drive T_2 into saturation, thus pulling the voltage at G to a low value. The output V_0 will then be pulled down to a value appropriate to logical 0 by T_4. It should be noted that, in the IC form, the separate transistors at the input are combined through separate emitter stripes on one base area (perhaps up to ten such stripes for this type of logic). Note should also be taken of the diode, which is a standard transistor but with a shorted collector–base junction. This type of IC is known as TTL (T^2L, or transistor–transistor logic) because it is two transistors (T_1, T_2) that are fundamental to the logic process.

The next example is a differential amplifier. In such an amplifier the output voltage V_0 is ideally linearly related to an 'inverting' input V_I and 'non-inverting' input V_N by $V_0 = G(V_N - V_I)$, where G is some gain (usually about 10^5 for an IC). In practice a small input value of $V_N - V_I$ is required

---

† This is a bond that occurs between metals at elevated temperatures (300–400 °C) when pressed together (gold-to-gold bonds are the easiest). Ultrasonic agitation is often used to help break through any oxide (as on aluminium) that may prevent good bonding.

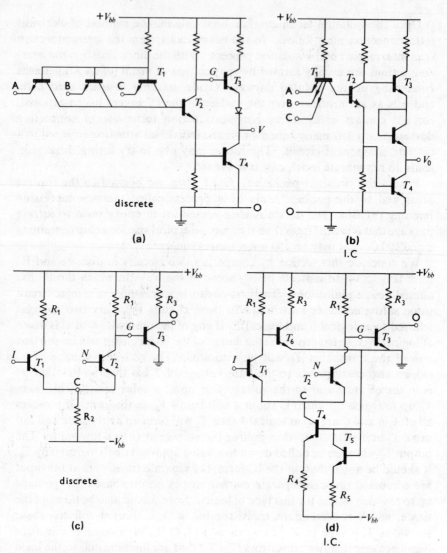

Fig. 5.14 Comparison of simple IC and discrete circuits. (*a, b*) NAND gate in TTL. (*c, d*) Input stages of differential amplifier.

to keep V_0 zero and this is referred to as the *off-set* voltage. In ICs the two sides of the differential amplifier can be kept balanced to a very high degree, thus keeping the off-set voltage low, but with very little extra cost. For example, the transistor T_6 is added in the IC to help preserve the symmetry of loading on the differential amplifier formed by T_1 and T_2. The cost of extra transistors is negligible in IC circuits. Although we shall not discuss the detailed operation of this well-known differential amplifier called the 'long-tail pair' it should be noted that in the discrete circuit the function of R_2 is to provide a constant current into the point C. In the IC one can tolerate many more components for a given additional cost. Thus T_5 is added to help stabilize the emitter current of T_4 against temperature variations and the consequent variations in V_{be}. The same change of V_{be} will occur to both T_4 and T_5 thus cancelling any net effect. The current through T_4 is then very well stabilized with temperature and the current into C is virtually constant. There will be additional components in the IC which will lower the output impedance and place the output voltage close to zero, but these are not shown here.

Detailed examples of the philosophy of IC design must be left for more specialized textbooks. Of course bipolar transistors and diodes are not the only devices that can be formed into ICs. Many other devices will be discussed later, and indeed any process where the yield is high and which does not consume large areas of silicon can be incorporated fairly readily. Although silicon is the preferred material at present, the control of epitaxial techniques with gallium arsenide is becoming very well developed, so that epitaxy can be used instead of diffusion.† Thus in the future GaAs may offer ICs with certain advantages for special applications.

5.8   The junction-gate field-effect transistor (JUGFET)

In many applications one would like to have an amplifier that accepts changes of voltage at its input but requires negligible changes of input current; in other words an amplifier having a high input impedance. Such a device is shown in Fig. 5.15; the input is a voltage that provides a varying

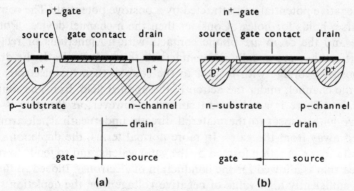

Fig. 5.15 Junction-gate field-effect transistors (schematic construction and circuit symbols).

† Arsenic tends to boil off the surface of GaAs in a diffusion furnace and makes the process difficult to control closely.

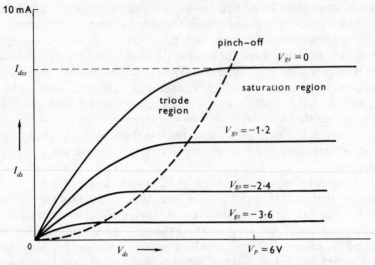

Fig. 5.16 FET characteristics.

reverse bias across a p-n junction and because of the reverse bias the input
takes a negligible current. This type of device is a *field-effect transistor*
(FET) though sometimes the full title shown at the top of this section is
used in order to distinguish it from other types of FET to be described later.
As for the transistor, there are two possible polarities: these are called
n-channel or *p-channel* to indicate which type of carrier transports the current.
Both types are shown with their circuit symbols in Fig. 5.15.

The physical action is straightforward and the correct bias arrangement
is easy to understand provided that one remembers electrons are repelled
by a negative potential and attracted by a positive potential. The converse,
of course, holds for holes. Consider then the n-channel device. Both the
source and the drain are ohmic contacts with no junctions or rectifying
properties (they are usually n^+ contacts to the n-channel). If the drain is
at a positive voltage with respect to the source, then the electrons move
along the channel, under the action of the applied field, from the source to
the drain. There is no diffusion current. However, as the gate is made
negative with respect to the material directly underneath it, electrons are
repelled away from the gate. In more normal terms, the depletion region
(around the p-n junction formed by the gate) widens, and so tends to reduce
the area that is allowed for the conduction of electrons. Indeed, if the gate
is at a sufficiently high value of negative voltage then the depletion region
extends right across the channel, and thereby severely restricts the current
flow. In fact, one finds that the current is not cut off but tends to a nearly
constant value. The analysis of the current flow in this saturation region is

difficult and is not attempted here. However, in the region (the triode region) below the pinch-off value the I-V characteristics (Fig. 5.16) are readily obtained and prove instructive.

The characteristics

The geometry and symbols given in Fig. 5.17 are used. It can be seen that an idealized one-sided structure is assumed, with the depletion only extending from the p^+ contact into the n-channel. The change in axial voltage is assumed to be sufficiently slow for the depletion region to support a total voltage of $[V(x) + \delta V - V_{gs}]$ at the point x if the gate has a voltage V_{gs} which is usually *negative* for a n-channel device. The diffusion potential for the p^+-n gate junction is δV as in previous work. The depletion region width is then found from eqn 4.5a and is given by

$$d(x) = [V(x) + \delta V - V_{gs}]^{1/2} d_0 / V_p^{1/2} \qquad 5.25$$

where d_0 is the thickness of the n-channel and V_p is the total voltage supported across a fully depleted channel (see Problem 5.4). Now the local field for driving the electrons is $-\partial V/\partial x$, so that the conduction current at the point x is given by the available area times the conduction current density:

$$I = eN_d\mu_n \frac{\partial V}{\partial x}(d_0 - d)W \qquad 5.26$$

This current is, of course, independent of x because there is no recombination or diffusion. The mathematics is made more elegant, and hence more easily followed, by writing $\phi = V(x)/V_p, \phi_g = (V_{gs} - \delta V)/V_p, eN_dWd_0\mu_nV_p/L = 3I_{dso}$, $X = x/L$. This gives, from eqn 5.26,

$$I_{ds}/I_{dso} = 3[1 - (\phi - \phi_g)^{1/2}] \, d\phi/dX$$

which can be integrated from $\phi = 0$ at $X = 0$ to $\phi = \phi_d = (V_{ds}/V_p)$ at $X = 1$:

$$I_{ds}/I_{dso} = 3\phi_d - 2(\phi_d - \phi_g)^{3/2} + 2(-\phi_g)^{3/2} \qquad 5.27$$

provided $\phi_g < 0$ and $[\phi_d - \phi_g] < 1$.

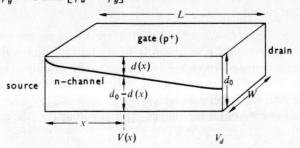

Fig. 5.17   FET idealized structure for analysis.

One can see that in this normalized form the *basic* parameters are I_{dso} and V_p and, given these, one can draw a universal set of characteristics. However, it must be noted that one does not usually bias the gate in the forward direction and I_{dss} is taken as the limiting saturated current with $V_{gs} = 0$. The value of δV, the diffusion potential, thus plays a part in modifying the universal characteristics and an appropriate set is shown in Fig. 5.16, with $\delta V = 0.6$ V and $V_p = 6$ V with $I_{dso} = 10$ mA. It should be noted that $\partial I_{ds}/\partial V_{ds} = 0$ whenever $V_{ds} + \delta V - V_{gs} = V_p$.† This is known as the *pinch-off* condition and it implies that the depletion extends right across the channel. For values of V_{ds} in excess of $[V_p + V_{gs} - \delta V]$ to the right of the dashed line in Fig. 5.16 different considerations must apply.

Suppose that $[V_{ds} - V_p - V_{gs} + \delta V] = [V_{ds} - V(x_0)] > 0$. Then the pinch-off point will occur at $x = x_0 < L$. The excess potential of V_{ds} over that required for pinch-off will then appear across an extended depletion region between the point $x = x_0$ and the drain. As V_{ds} increases so the point x_0 moves away from the drain. This extended depletion will not in itself significantly alter the drain current because the sense of the fields across the depletion region is to drive the electrons rapidly across the depletion to the drain, in spite of a 'depletion'. However, there will be a shortening of the undepleted channel as x_0 moves towards the source. The resistance of this part of the channel will *fall* although, at constant gate voltage, the voltage across this part will *remain* close to $|V_p + V_{gs} - \delta V|$. The current will then tend to rise. A calculation of the effect of this *channel length modulation* for the idealized structure shown in Fig. 5.17 is in fact misleading, because in practical FETs the drain is not directly adjacent to the gate. The depletion region from the pinch-off point x_0 now extends beyond the edge of the gate towards the drain. Changes in this direction then help to absorb the excess potential of V_{ds} and reduce channel length modulation *under* the gate. Experimentally it is found that the current can nearly saturate at constant V_{gs} as V_{ds} is increased—as shown in the idealized *saturation region* of Fig. 5.16.

The equivalent circuit

The input current is so low in the FET that the *h*-parameters are not really appropriate; they would give absurdly high and uncertain values to the forward-current transfer ratio h_f. For this reason the output-current generator is proportional to the input voltage rather than the input current (Fig. 5.18a). One refers to a *mutual conductance*,‡ $g_m = \delta I_{ds}/\delta V_{gs}$, as giving the change of output current for the change of input voltage. At the condition of

† $\partial I_{ds}/\partial \phi_d = 0$ if $\phi_d = 1$.
‡ Also called transconductance.

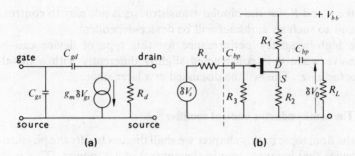

Fig. 5.18 Circuits for FET. (*a*) Equivalent circuit (at low frequency). (*b*) Simple amplifier. Note alternative symbol for device.

pinch off the reader will be able to show by differentiation that

$$g_m = (3I_{dso}/V_p)\{1 - [(\delta V - V_{gs})/V_p]^{1/2}\}$$ 5.28

The output impedance (R_d in Fig. 5.18a) cannot be estimated from this simple theory and must be left for experimental determination; it is usually high enough to be neglected in many practical circuits. At modest frequencies it should be remembered that there are capacitances between the gate and source and between gate and drain, determined essentially from the geometry, bearing in mind the depletion region. These values are typically a few picofarads. There is no diffusion capacitance, because the gate-channel junction is on reverse bias.

The design of the amplifier (Fig. 5.18b) follows closely the considerations outlined for the bipolar amplifier (Fig. 5.8a). One must consider breakdown, power dissipation, load current, and operating point for adequate variations in voltage across the device. The bias circuit appears superficially similar to that for the bipolar transistor but now the gate-source junction is on *reverse* bias, so that in the corresponding n-channel device the bias voltage V_{gs} is derived across resistor R_2 by the current I_{ds} (($V_{gs} = -I_d R_2 = -kV_p$). The gate is held in the steady state at 0 volts by the resistor R_3. Because the gate takes negligible current, R_3 can be really large ($R_3 \sim 10$ M ohms).

The circuit analysis again follows that for the bipolar device in many respects. The supply is always considered to be a short circuit as far as *changes* in input and output are concerned. The input impedance is limited by R_3 [and any input capacitances]. Redrawing the circuit, and neglecting ineffective components, is the key to understanding. Thus, ignoring the bypass capacitances and taking $R_d \gg R_1, R_L \gg R_1, g_m R_1 \gg 1$ and $g_m R_2 \gg 1$, the low-frequency gain of voltage is given by $V_0/V_s \simeq -R_1/R_2$ independent of the precise device parameters. If, however, R_2 is bypassed with a large-enough capacitor then one can show that the gain is approximately $-g_m R_1$

but, like β and R_e for the bipolar transistor, g_m is not easy to control for the FET and so such an amplifier will be device-dependent.

The high-frequency performance for this type of device can be very impressive when it is constructed slightly differently with a metal–semi-conductor gate. This will be discussed at a later stage.

5.9 The semiconductor control rectifier

As the final topic in this chapter we shall discuss briefly the physical action of a device that is much used in the control of a.c. power. This is a switch formed by adding yet another layer to a transistor-type structure. The class of layered n-p-n-p, n-p-n-p-n-p etc. devices are called *thyristors;* however, space will only permit us to discuss one of these—the semiconductor control rectifier (SCR).

Figure 5.19 shows schematically the construction of a medium-power silicon n^{++}-p^{+}-n-p^{++} SCR and its appropriate circuit symbol. In principle

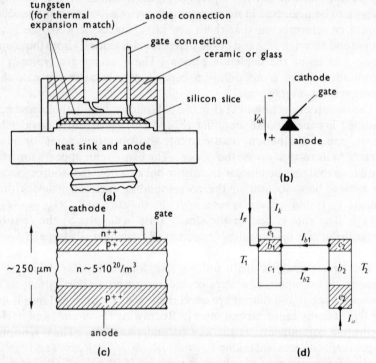

Fig. 5.19 The semiconductor (silicon) control rectifier. (*a*) Construction of medium power device (schematic). (*b*) Circuit symbol popularly used. (*c*) Cross-section of silicon slice. (*d*) Two-transistor approximation.

there is also the complementary device, but we shall not consider this. Although the SCR has four layers, it generally has only three electrodes—the cathode, gate and anode. It should be noted that the n^{++}-p^{+}-n layers are actually constructed, and indeed behave, like a n-p-n bipolar a.c. power transistor. The dimensions are much larger than those of the idealized high-performance transistor considered previously, but then the device only needs to work at low frequencies. In this SCR the cathode is the emitter and the gate is the base, but the collector is only contacted through the final p^{++}-layer. This final layer can be considered to be the emitter of another transistor—a p-n-p transistor—which has a rather poor base-transport factor because of a very long, lightly doped, n-type base. This second transistor therefore has a poor α factor. A ready understanding of the SCR can be based on its approximate equivalence to two transistors. The arguments are presented below.

One finds that in the absence of any gate current the current between the cathode and anode is blocked in both directions of applied voltage; one or the other of the p^{+}-n junctions is on reverse bias so that the voltage can be supported mainly by a depletion in a long, relatively lightly doped, n-region (as in a high-voltage rectifier). As in a rectifier, there will be some breakdown voltage V_b (though we shall see that this value is at a lower voltage than the corresponding simple rectifier). Now consider what will happen if the anode is at a positive voltage with respect to the cathode, so that both transistors in Fig. 5.19d can be considered to be active. Then suppose that into the middle p^{+}-region or gate is fed some current I_g. From current continuity it will follow that $I_a = I_k - I_g$ (notation as in the figure). Now the base current of T_2 is a current of electrons comprising $\alpha_1 I_k$ from the n^{++}-emitter and a reverse saturation current of electrons I_{co2} from the collector of T_2. These electrons control the hole charge in the base of T_2. Equally the current flowing into the base of T_1 is made up from $\alpha_2 I_a$ of hole current from the emitter of T_2 and a reverse saturation current I_{co1} of holes (minority carriers) from the collector of T_1. The gate current has been included separately in this balance to give a continuity condition:

$$I_k - I_g = I_{b1} + I_{b2} = \alpha_1 I_k + \alpha_2 I_a + I_{co1} + I_{co2}$$

$$\text{or} \qquad I_a = (\alpha_1 I_g + I_{co1} + I_{co2})/(1 - \alpha_1 - \alpha_2)$$

5.29

As soon as $\alpha_1 + \alpha_2$ becomes unity the current is no longer controlled by the gate but by the external circuit, and eqn 5.29 breaks down. Indeed the numbers of electrons and holes in the long n-type region that has supported the large voltage now become so great that the device switches to a low-impedance high-current state.† However, the reader may well be mystified

† The two equivalent transistors can be said to be saturated.

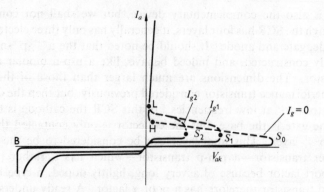

Fig. 5.20 SCR characteristics (schematic). S_0—switching point at breakdown for forward bias. S_1, S_2—switching points for gate currents I_{g1} and $I_{g2}(I_{g2} > I_{g1} > 0)$. L—latching point. H—holding point. B—reverse breakdown with no gate current.

at this point because in previous work the α parameter has been taken as a constant varying very little with current. At very low currents, however, other factors can affect the value of α and reduce it in magnitude. The simplest effect is leakage from the emitter to the base, perhaps caused by a surface current at the junction. If this leakage has a resistance R_s then, as soon as the emitter resistance $R_e \sim kT/eI_e$ rises to become comparable to R_s, the emitter efficiency will fall and α will be lowered. One also finds that recombination can vary, with current level complicating the theory still further. Moreover there is one reasonable transistor and one poor transistor, so that at low currents it is found that $\alpha_1 + \alpha_2$ is less than unity. However, if the gate current raises the current level, then the α parameters can increase sufficiently to allow switching into the low-impedance high-current state. One of the great virtues about the SCR is that the gate current is not needed continuously; once the low-impedance state is initiated it sustains itself and the gate current can be removed. The device only reverts to its high-impedance state by the main current falling to a low value.

Figure 5.20 indicates some of the important features of the switching characteristic of the SCR. One notes that as soon as the combination of gate current and applied voltage move the SCR into the *switching point* (one such point for every value of I_g) then the device becomes unstable and switches to a current primarily determined by the external circuit. The gate current must be applied until the current exceeds the *latching point*. If the current is removed before this point the SCR will switch back into the off state. In general, one has to be careful about the rate of build up of the total device current. The low-impedance state is initiated around the gate contact and has to spread to the whole device area before the full rated current may be taken. If the current is allowed to build up too soon the device can burn out. To switch the device back into the high-impedance state the device current has to be lowered below the *holding point*, when once again it becomes unstable and switches. Note that in the unstable region the differential conductance $\partial I_a/\partial V_{ak}$ is negative. It was seen how a negative conductance could lead to switching for the tunnel diode; this is a similar principle being operated here (though of course there is no suggestion of tunnelling).

In operation there are other methods of initiating switching into a low-impedance state than the simple application of a gate current. A high rate of change of anode–cathode voltage can change the depletion width in the lightly doped n-region, so that electrons are pushed through the device fast enough to raise the current to the switching level. Thus there is a maximum limit to the value of $\partial V_{ak}/\partial t$. Incident light can liberate electrons and cause switching and, this can be used in a special device called the light activated SCR or LASCR. Temperature rises can increase the reverse saturation currents to a level where switching is initiated. In this context it is also as well to mention the effects of breakdown at high fields. We shall learn (Chapter 8) how high electric fields can accelerate electrons and holes to sufficient energies for them to liberate more electrons and holes from the crystal lattice. Thus at high fields an injected current I_0 can be multiplied up to appear at the terminals as MI_0, with $M > 1$. All the currents entering the high-field, lightly-doped, n-region in the SCR under discussion will be multiplied by some factor, say M, and so switching can occur at $1 - M\alpha_1 - M\alpha_2 = 0$. We shall see later that normal breakdown requires M to become infinite, and so normal breakdown occurs at higher fields than the SCR switching.

Figure 5.21 shows one of the simplest and most common uses for an SCR

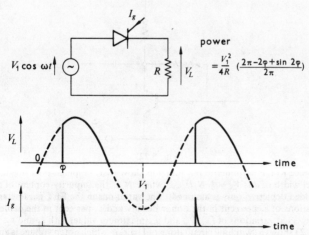

Fig. 5.21 Simple phase-control use of SCR (trigger source and timing circuit not shown).

in controlling a.c. power. This is *phase control*. The gate is switched on at some controllable phase angle ϕ and initiates rectification of the a.c. Once the anode current has fallen to a low value the SCR reverts to its high-impedance state. Thus the amount of power delivered to the load from the mains supply can be controlled by the phase of initiation of rectification. More complicated circuits are generally used, with inductors to smooth the current flow. Small devices called Triacs, which have extra p- and n-layers and behave like two SCRs back to back, are often used in domestic light dimmers. This is also a phase-control application. It should not be thought that this is the only application, for these versatile switches can be used as pulse modulators, d.c. controllers, d.c.-to-a.c. inverters and vice versa, among other applications. The reader should consult reference 5.9.

PROBLEMS

5.1 Explain why I_{eh}/I_e will be larger than N_{ao}/N_{de} (using the notation of section 5.2). The explanation should consider the likely rates of recombination in the differently doped regions compared to each other, and then look at the more detailed expressions for the components of current carried by the electrons and holes in an asymmetric diode (eqns 4.11a and 4.11b).

5.2 Redraw Fig. 5.7a with the emitter common to both sides of the circuit. Short the emitter–collector to find the collector current when a base current I_b flows (hence find h_{fe}) and also determine the effective emitter–base resistance h_{ie}, remembering that now both base and collector current flow in h_{fb}. Then open-circuit the base and find the parameters h_{re} and h_{oe} on the assumptions $h_{rb} \sim O(1/h_{oe}) \gg h_{fe}R_e \gg R_e$. Hence establish eqn 5.5.

5.3 Figure 5.22 indicates an alternative method of biasing a transistor. Show that, for minimum variations of power dissipation with variations in β, $\beta_0 R_1 = R_2$ is a good choice, where β_0 is the mid-value of the permissible range of values for a given type of transistor. Show also that, although β may vary by a factor of 4:1, the emitter current varies by no more than a factor of 2. (β is large.)

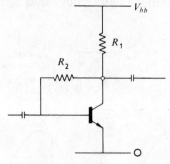

Fig. 5.22

5.4 By reference to the asymmetric p-n junction show that the pinch-off voltage in the FET with a channel width of d is $V_p = \frac{1}{2}eN_id^2/\varepsilon_0\varepsilon_r$ where N_i is the impurity content of the channel and a single-sided depletion only is assumed. The variations in the FET parameters arise as a result of variations of ε_1 per cent in the donor density and ε_2 per cent in the depletion width. Discuss the expected variations of I_{dss}, V_p, and g_m (maximum value) with ε_1 and ε_2.

5.5 Figure 5.23 shows a switching circuit (long tail pair) in which a step voltage is applied to the base of T_1. Explain why the switching speed can be much greater than that of the circuit shown

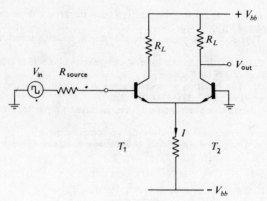

Fig. 5.23

in Fig. 5.12. With the use of the simple high-frequency circuit discuss the limitations on this switching speed. $[IR_L < (V_{bb} - 2)$ so that saturation does not occur.]

5.6 For the circuit of Fig. 5.21, verify that the power delivered to the load R is $(V_1^2/8\pi R) \times (2\pi - 2\phi + \sin 2\phi)$. Sketch this power as a function of ϕ.

5.7 The admittance parameters are given by

$$I_1 = Y_{11}V_1 + Y_{12}V_2, \qquad I_2 = Y_{21}V_1 + Y_{22}V_2$$

Sketch an equivalent circuit with the appropriate admittances and current generators. Compare it with the equivalent circuit for the FET given in Fig. 5.18a and show that $Y_{11} = j\omega(C_{gs} + C_{gd})$, $Y_{12} = j\omega C_{gd}$, $Y_{21} = g_m$, $Y_{22} = G + j\omega C_{gd}$ where $G = 1/R_d$.

5.8 Discuss the effects of eqn 4.5b on the design of a transistor collector region, bearing in mind the discussion below eqn 5.23. What are the conflicting requirements? Show from considerations of transit time and the breakdown field (E_b) that the peak-to-peak r.f. voltage is the order of $E_b v_s/\omega_{max}$ for useful operation. The scattering limited velocities are taken as v_s. If the breakdown field is taken to be independent of the dimensions of any transistor show that for transistors, which are designed to operate at a succession of frequencies, the useful peak-to-peak voltage will decrease as $1/f$. Hence show that, for a constant impedance load, the power will decrease as $1/f^2$. Allowing for the maximum power transfer theorem, make an estimate of the power x impedance x frequency product ($E_b \sim 350 \times 10^5$ V/m). [Answer: around 10^4 W ohm GHz^2 but this can be reduced by the shunting effect of the capacitive impedance of the collector depletion.]

5.9 Compare the admittance parameters given in Problem 5.7 with the h-parameters, using a similar notation. In particular show that $Y_{22} = h_{22} - (h_{12} . h_{21}/h_{11})$.

Hence show that the output impedance $1/Y_{22}$ of the common emitter mode of operation for a bipolar transistor with $\delta V_{be} = 0$ is approximately β times higher than the output impedance $1/h_{22}$ also in the common emitter mode but with $\delta i_b = 0$.

General references

LINDMAYER, J. and WRIGLEY, C. Y. Reference 1.2.

BECK, A. H. W. and AHMED, H. Reference 1.1.

Special references

Transistors

5.1 EBERS, J. J. and MOLL, J. L. Large signal behaviour of junction transistors. *Proc. IRE*, **42** (1954), 1761–72.

5.2 BEAUFOY, R. and SPARKES, J. J. The junction transistor as a charge controlled device. *ATE J.* (*London*) **13** (1957), 310–324.

5.3 EARLY, J. M. Effects of space-charge layer widening in junction transistors. *Proc. IRE*, **40** (1952), 1401–06.

5.4 GUMMEL, H. K. and POON, H. C. An Integral charge control model of bipolar transistors. *Bell System Tech. J.*, **49** (1970), 827–851.

5.5 WEBSTER, W. M. The variation of junction transistor current gain amplifier factor with emitter current. *Proc. IRE*, **42** (1954), 914–920.

5.6 WATSON, H. A. *Microwave Semiconductor Devices and their Circuit Applications.* McGraw-Hill, 1969.
    Good section on microwave transistor design compromises.

5.7 DELHOLM, L. A. *Design and Application of Transistor Switching Circuits.* McGraw-Hill, 1961.

Field effect transistors

5.8 SEVIN, L. J. *Field Effect Transistors.* McGraw-Hill, 1965.

Silicon control rectifiers

5.9 GENTRY, F. E., GUTZWEILER, F. W., HOLONYAK, N., and VON ZASTROW, E. E. *Semiconductor Controlled Rectifiers.* Prentice Hall, 1964.

Integrated circuits

5.10 LYNN, D. K., MAYER, C. S., and HAMILTON, D. J. *Analysis and Design of Integrated Circuits.* McGraw-Hill, 1967.

5.11 HIBBERD, R. G. *Integrated Circuit Pocket Book.* Newnes-Butterworth, 1972.

5.12 *Integrated Electronic Systems.* Reference 1.7.

Analysis and design of circuits

5.13 ANGELO, E. J. *Electronics BJT, FETs and Microcircuits.* McGraw-Hill, 1969.

5.14 FITCHIN, F. C. *Transistor Circuit Analysis and Design.* 2nd edition. Van Nostrand, 1966.
    A very thorough consideration of stability criteria in transistor circuit bias arrangements.

5.15 KOHONEN, T. *Digital Circuits and Devices.* Prentice Hall, 1972.

5.16 PASCOE, R. D. *Solid State Switching*, Wiley, 1973.
    The General Electric series of manuals are also helpful, especially with practical advice.

6

Waves in Crystals

6.1 Waves and particles

The aim of this chapter is to explain the ideas of energy bands, effective mass and holes in a more rigorous manner than in Chapter 2, but using only the wave-like nature of the electron rather than any advanced quantum mechanical treatment. Thus the starting point, with which the reader is assumed to be familiar, is the dual relationship between particles (with total energy $\mathscr{E}$ and a momentum p) and waves (of wavelength λ and frequency f) expressed by the relationship $\lambda = h/p$ and $\mathscr{E} = hf$. The experimental evidence for this is well covered in textbooks on quantum theory.[2.3–2.8] It is more convenient to use here the angular frequency $\omega = 2\pi f$ and the propagation constant $k = 2\pi/\lambda$, so that a free electron is to be associated with the wave

$$\psi(x, t) = \psi_0 \exp j(\omega t - kx) \qquad 6.1a$$

where the expected values of the electron's energy and momentum are given by

$$\mathscr{E} = \hbar\omega \quad \text{and} \quad p = \hbar k \qquad (h = 2\pi\hbar) \qquad 6.1b$$

respectively. Conservation of energy still holds, so that

$$\mathscr{E} = \mathscr{E}_0 + (p^2/2m) \qquad 6.2$$

where $\mathscr{E}_0$ is the potential energy, assumed constant, and $(p^2/2m)$ gives the kinetic energy. It is not normal to talk about the frequency of electron waves because the zero reference of energy is not unique for the potential energy and so the 'frequency' ω cannot be unique. Thus, although one can formally write

$$\omega - \omega_0 = \hbar k^2/2m \qquad 6.3a$$

where $\mathscr{E}_0 = \hbar\omega_0$, it is more usual to consider the $\mathscr{E} - k$ relationship

$$\mathscr{E} - \mathscr{E}_0 = (\hbar k)^2/2m \qquad 6.3b$$

The ω-k relationship is the *dispersion* relationship for the wave. These results of course follow from Schrodinger's equation

$$(\hbar^2/2m)\frac{\partial^2\psi}{\partial x^2} + (\mathscr{E} - \mathscr{E}_0)\psi = 0 \qquad\qquad 6.4a$$

with the operator equations

$$-j\hbar\frac{\partial\psi}{\partial t} = \mathscr{E}\psi; \qquad j\hbar\frac{\partial\psi}{\partial x} = p\psi \qquad\qquad 6.4b$$

One asserts that the square of the wave's amplitude $|\psi|^2$ gives the probability of observation of the electron, or the density of the electrons if there are more than one to be considered. This is in accord with one's ideas on light waves, where the power is proportional to the square of the amplitude and the power would be expected to be proportional to the number of particles, *photons*, associated with the wave.

Consider now a hypothetical emission process in which the electrons are allowed out of some container once every τ seconds. The value of $|\psi|^2$ could be given by $|\psi|^2 = A^2 \cos^2 (2\pi t/\tau)$, so that for electrons of energy $\hbar\omega$:

$$\psi(0, t) = A \cos \omega_m t \exp j\omega t \qquad (\omega_m = 2\pi/\tau) \qquad \text{(Fig. 6.1)}$$

$$= \tfrac{1}{2}A[\exp j(\omega + \omega_m)t + \exp j(\omega - \omega_m)t] \qquad\qquad 6.5$$

Thus the attempt to localize the particle in time has led to more than one frequency in the wave (a *wavepacket*) and a resulting uncertainty in the energy of the electron as given by eqn 6.1. One notes that the localization in time Δt and the spread in energy $\Delta\mathscr{E}$ are linked because $\omega_m\tau = 2\pi$, giving

$$\Delta\mathscr{E} \cdot \Delta t \sim h \qquad\qquad 6.6$$

Fortunately the value of h is so small that such uncertainties are of little concern here (see problem 6.1). However the corresponding result for momentum and position is not so negligible. This is now considered.

As the electron moves away from the source, ψ changes. The change in frequency of ω to $\omega + \omega_m$ changes k to $k + k_m$ where, from a knowledge of the dispersion relationships, one can say $\omega_m \simeq (d\omega/dk)k_m$. Then one can determine $\psi(x, t)$ from eqn 6.5

$$\psi(x, t) = \tfrac{1}{2}A\{\exp j[(\omega + \omega_m)t - (k + k_m)x]$$

$$+ \exp j[(\omega - \omega_m)t - (k - k_m)x]\} \qquad 6.7a$$

$$|\psi|^2 = A^2 \cos^2 (\omega_m t - k_m x) \qquad\qquad 6.7b$$

Thus the electrons that were localized in time at $x = 0$ are also localized in space as they travel away from the source. The peak probability travels with a velocity, called the *group velocity*, given by $v_g = \omega_m/k_m = d\omega/dk$. This must be the velocity with which the particles are actually moving through space because the peak probability locates the expected position

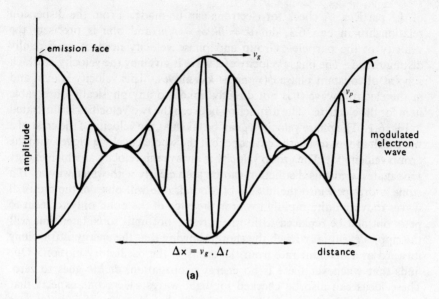

(a)

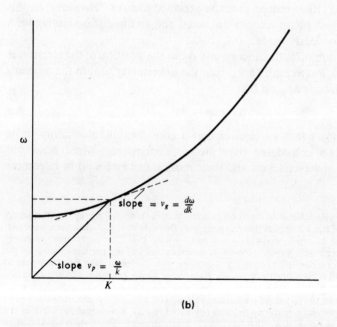

(b)

Fig. 6.1 Periodic emission process illustrating distinction between group and phase velocity. (a) v_g is velocity of envelope; v_p is velocity of crest inside envelope. (b) Group and phase velocity illustrated by different slopes on dispersion diagram.

of the particle. A check for electrons can be made. From the dispersion relationship in eqn 6.3, $d\omega/dk = \hbar k/m = p/m$ and p/m is precisely the velocity of the particle. Group and phase velocity need to be carefully distinguished. The phase velocity of a wave is given by the velocity at which a point of constant phase $\phi(=\omega t - kx)$ moves. This velocity is ω/k, and in the electron wave it is not directly linked to any physically observable item for the electron. The distinction between the two velocities is illustrated in Fig. 6.1. The phase velocity v_p can be taken as the velocity of the crest of a wave as it moves inside the envelope, but the whole envelope moves with the group velocity v_g. The group velocity is always the velocity at which energy propagates, and this distinction can be seen clearly with water waves. If a stone is dropped into the middle of a pond, one will observe the crests of waves moving quite rapidly towards the edge of the pond only to seem to peter out and be replaced with other crests; not until some later time will the ripples eventually reach the shore, showing that the energy is travelling forward at a different rate from the crests of the oscillatory motion. One finds that whenever there is no energy propagation, $d\omega/dk$ goes to zero. These ideas can also be checked for light waves. Here one expects that photons would travel at the velocity of light, *in vacuo*. So $d\omega/dk = c$ (because $\omega = kc$) does indeed give the expected answer. However, in this case the group and phase velocity are equal and so the distinction between the two is not so clearly made.

Returning to eqn 6.7b, the uncertainty Δx in the position of the electron is approximately a wavelength $2\pi/k_m$: but the uncertainty, Δp, in the momentum of the electron is $\hbar k_m$, so that

$$\Delta x \, . \, \Delta p \sim h \qquad\qquad 6.8$$

Problem 6.1 shows that an electron has a considerable uncertainty in its momentum when considered over atomic dimensions. More advanced considerations of wavepackets and their motion can be found in references 2.5 and 2.6.

Finally it can be noted that if the total energy is less than the potential energy then the dispersion relationship of eqn 6.3 implies that k is imaginary (because $\mathscr{E} < \mathscr{E}_0$); then the spatial part of the wave varies as $\exp(-ax)$ with $a = [2m(\mathscr{E}_0 - \mathscr{E})/\hbar^2]^{1/2}$. The momentum is clearly undefined, but there is still a finite probability (given by $|\psi|^2$) of the particle existing on the 'wrong' side of the barrier, this probability decreasing with the penetration into the potential barrier. Such a 'wave' is termed *evanescent*. Evanescent waves are the key to the quantum mechanical tunnelling effect. With very thin potential barriers there is a finite probability of an electron appearing on the far side of the barrier, although on a classical argument the potential barrier is too high for any penetration to occur (see Problem 6.2). This is then the basis for the tunnel process in the tunnel diode that has already been discussed. There are other devices in which tunnelling can be important. Electrons can tunnel through thin insulating films between metals (thickness of films around 50 Å) and can also tunnel across the band gap at high-enough electric fields. This latter process is the Zener effect (see page 184).

6.2 Waves in periodic structures

One of the major advances in the theory of conduction has been the analysis of electron-wave propagation through periodic arrays of atoms. The modern theory, while complex in detail, is simple and elegant in outline. The electrons which contribute to the conduction processes are those which are almost free of their parent atoms and so should have wave motions not unlike those of free electrons. However, near the nucleus of each atom the electric potential is varying by large amounts over short distances. Consequently any free electron wave with a given energy has its propagation constant changing rapidly around the atoms in the crystal. For a long time this behaviour of the wave function around the core of the atom made it difficult to compare the crystal waves with free electron waves. However, the important aspects of the potential variation around the nucleus are those affecting its ability to scatter plane waves; the local details of the potential variations are of less interest. As a crude analogy one may say that the motion of waves in the sea around an array of rocks is not necessarily described by the detailed shape and size of each rock but rather by the scattering properties of the rock to the incoming waves. Each rock could be replaced by an idealized rock which had the same wave scattering properties, but the shapes of the ideal and real rocks could be quite different. So in modern theory of electron waves in a crystal the true potential of the atom is replaced with a 'pseudo-potential' which is designed to give the correct scattering of a plane electron wave. Relatively simple forms of 'pseudo-potential' may be used, so that the difficulties of computing real problems, of matching experimental results with computed data and of predicting new results are alleviated. The method has been highly successful in furthering the detailed understanding of conduction processes. It also is the justification for an elegant tutorial model given by V. Heine[6.3] to explain how the electronic band structure arises in a periodic array of atoms. It is Heine's tutorial model which we now discuss.

We consider the wave motion of a plane wave in the direction of a wave-vector $\mathbf{k}$; this is a vector whose magnitude, $|\mathbf{k}|$, equals the propagation constant $2\pi/\lambda$. The crystal is a three-dimensional array of atoms, and along the direction of $\mathbf{k}$ it is supposed that the pattern repeats in a distance $L(\mathbf{k})$. For convenience the axes are chosen with $\mathbf{k}$ in the $0x$ direction. A free electron wave of energy $\mathscr{E}$ with respect to the reference energy $\mathscr{E}_0$ is incident on one of the unit cells of the crystal. It will be supposed that the variations of potential inside the cell combine to reflect a fraction R of an incident wave and transmit a fraction T. The phase factors of these reflection and transmission coefficients are adjusted to the plane at the edge of the cell even though the reflection is distributed throughout the cell. Symmetry will be assumed, so that it will not matter if the wave is incident from the left or the right of the cell. For a plane wave $\psi_i \exp j(\omega t - kx)$ incident on the plane $x = 0$ (Fig. 6.2a), a wave $R\psi_i \exp j(\omega t + kx)$ is reflected, while $T\psi_i \exp j(\omega t - kx)$ is transmitted to the next cell. The phase of ψ_i is immaterial in the argument at present so ψ_i is taken as real. The numbers of electrons must be conserved, so that the probable number that are incident must

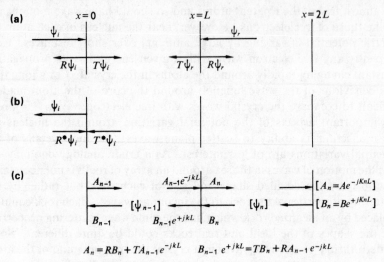

$$A_n = RB_n + TA_{n-1}e^{-jkL} \qquad B_{n-1}\,e^{+jkL} = TB_n + RA_{n-1}\,e^{-jkL}$$

Fig. 6.2 Scattering from cell walls in periodic crystal.

equal the probable number reflected plus the probable number transmitted, thus

$$TT^* + RR^* = 1 \qquad\qquad 6.9$$

(* denoting complex conjugate so $|R^2| = RR^*$)

Now if $S(t)$ is a time-varying solution to a physical problem then $S(-t)$ is another realizable solution, though perhaps of a slightly different problem, provided that energy is conserved for either example. Energy is conserved for the motion of an electron in an electrostatic potential, so that this process can be used in our discussion. If time is reversed, and also the complex conjugate is taken, one obtains another solution in which the original transmitted wave becomes an incident wave $T^*\psi_i \exp j(\omega t + kx)$ coming from the right rather than the left (Fig. 6.2b). Similarly, the original reflected wave becomes another incident wave, from the left rather than the right, $R^*\psi_i \exp j(\omega t - kx)$. These two waves now combine at the plane $x = 0$ to produce the outgoing wave $\psi_i \exp j(\omega t + kx)$. Thus the reflected part of $T^*\psi_i$ and the transmitted part of $R^*\psi_i$ must cancel, hence

$$TR^* + RT^* = 0 \quad \text{or} \quad R/T = -R^*/T^* \qquad\qquad 6.10$$

We now look for a pattern of reflected and transmitted electron waves such that the probability of an electron being found at any point within each cell of the crystal is identical from cell to cell (remember that the crystal is perfectly periodic). Thus the pattern of the free electron waves will be $[A \exp j(\omega t - kx) + B \exp j(\omega t + kx)]$, but it must be remembered also that there could be an arbitrary phase factor, say $\exp -jKL$, in front of this expression without alteration of $|\psi|^2$. Provided that the phase factor changed by a constant amount from cell to cell, the periodicity would not be violated in any sense. Thus solutions are sought so that in the nth cell the wave function is given by

$$\psi_n = (A \exp -jkx + B \exp jkx)(\exp j\omega t) \exp -jnKL \quad (0 < x < L) \quad 6.11$$

The wave function is continuous at the boundaries, so that the reflection and transmission coefficients can be used to match up the amplitudes from cell to cell. This gives (Fig. 6.2c):

$$TA \exp \left[-j(n - 1)KL\right] \exp - jkL + RB \exp - jnKL = A \exp - jnKL$$

$$TB \exp - jnKL + RA \exp \left[-j(n - 1)KL\right] \exp - jkL$$
$$= B \exp -j(n - 1)KL \exp jkL$$

Writing $T' = T \exp jkL$ and $R' = R \exp jkL$ one can rearrange the above to

$$T' \exp 2jKL - (T'^2 + 1 - R'^2) \exp jKL + T' = 0 \qquad 6.12$$

It is then noted that T' and R' satisfy eqns 6.9 and 6.10, so that

$$(T'^2 + 1 - R'^2)/T' = T' + \frac{1}{T'} + \frac{R'R'^*}{T'^*} = \left(\frac{1}{T'} + \frac{1}{T'^*}\right) = \frac{2}{T_0} \cos (kL + \phi)$$

assuming that $T = T_0 \exp j\phi$. Then from eqn 6.12

$$T_0 \cos KL = \cos (kL + \phi) \qquad 6.13$$

The value of the energy $\mathscr{E}$ gives the value of k from eqn 6.3 with $k = (2m\mathscr{E})^{1/2}/\hbar$. Equation 6.13 then yields the $\mathscr{E} - K$ relationship for the electron waves in a periodic structure. The phase shift per unit cell is considered to give the net propagation constant K of the wavepacket of free electron waves formed by the mixture of forward and reverse free electron waves.

Before discussing the nature of the solutions to eqn 6.13 it must be emphasized that it was no happy accident which enabled a solution of the form of eqn 6.11 to be evaluated. A powerful mathematical theorem called Bloch's theorem, or Floquet's theorem, states that such a solution will exist in a periodic system of the type considered. Discussions of this theorem can be found in references 2.1 and 6.5.

132

The Brillouin diagram

The propagation of the electron through the crystal requires that, in eqn 6.13, K shall be real. If K is imaginary then the probability of an incident electron wave penetrating into several cells of the crystal will be an exponentially decaying function. In the bulk of the semiconductor this probability will be negligible. Hence imaginary values of K cannot correspond to electron states in the bulk of the crystal. The approach shows that there is a whole range of energy bands with propagating waves. This feature can most easily be demonstrated by putting $\phi = 0$ in eqn 6.13. The energy and k are related as above so that eqn 6.13 may be re-written as

$$\cos KL = T_0^{-1} \cos (\mathscr{E}/\mathscr{E}_L)^{1/2} \qquad\qquad 6.14$$

where $\mathscr{E}_L = \hbar^2/2mL^2$. Propagation is only possible when the right-hand side of eqn 6.14 is less than unity. However $1/T_0$ is fundamentally larger than unity because a part of the wave amplitude is reflected. There are then energies centred around $(\mathscr{E}/\mathscr{E}_L)^{1/2} = n\pi$ for which propagation is not allowed (Fig. 6.3a). One may note how close to the free-electron waves are the solutions, except for the forbidden energy regions.

The dispersion relationship of eqn 6.14 differs in another important respect from that for free electrons; it is periodic in KL. For example, if an electron in the crystal were to have its momentum altered by $\hbar\, \Delta K = \hbar k_0 + n\hbar(2\pi/L)$ then its physical velocity, which is related to the group velocity or $d\mathscr{E}/dK$, is changed by the same amount regardless of the value of the integer n. The interpretation of this is that the momentum $n\hbar(2\pi/L)$ goes to the crystal as a whole. Processes in which this occurs are termed *umklapp* processes; these must be left for further reading.[2.1] Consequently all the electron states can be defined within a range $-\pi < KL < \pi$. This is the *reduced zone* scheme (Fig. 6.3b) and it gives the most usual method of presenting the *Brillouin diagram* as the $\mathscr{E} - K$ relationship is called (after L. Brillouin).

The Brillouin diagram is really a four-dimensional relationship: $\mathscr{E}$ with the three vector directions of $\mathbf{K}$. The reduced zone is then a volume, the *Brillouin zone*, whose boundaries are defined by the wave-vectors $\mathbf{K}$ such that $|\mathbf{K}| = \pi/L(\mathbf{K})$, where $L(\mathbf{K})$ is the repetition length of the cell, in the direction of $\mathbf{K}$. The space defined by all K-vectors is called K-*space* or *reciprocal space;* its dimensions are those of reciprocal length. In the simple $\mathscr{E} - K$ relationship of eqn 6.14, where the transmission factor T_0 has been taken as a constant, the extrema always occur at the edges or the centre of the Brillouin zone. This simple result can be understood by considering the constructive reflections from a series of obstacles spaced one half-wavelength apart in a long line. There is then always a whole

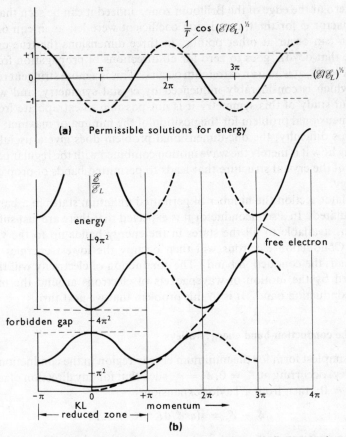

(a) Permissible solutions for energy

(b)

Fig. 6.3 Brillouin diagram (for one-dimensional periodic structure).

number of wavelengths between the reflection from the initial obstacle and the reflection from the nth obstacle—allowing for the trip forward and back. This physics of reflection can be placed on a firmer footing as in Problem 6.3. Unfortunately it gives a misleading result for three-dimensional propagation, and for other practical cases where the transmission factor is a function of wavelength and direction. In these more practical examples the derivative of a branch can be non-zero at the edges of the Brillouin zone at certain critical directions where there is actually no reflection from the edge of the Brillouin zone in that particular direction. The derivative of the branch may may then be a $d\mathscr{E}/dK = +a$, for example, at such a point. However the periodic nature of the problem is found to imply that there is another branch through the same critical point, but with a derivative $d\mathscr{E}/dK = -a$. Thus in the more realistic problems one finds that $d\mathscr{E}/dK$ does not necessarily go to zero at the edge of the Brillouin zone; indeed it can be seen that if the phase factor ϕ for the transmission coefficient were left as in eqn 6.13 the extrema can occur at other points. In three dimensions the true extrema require that $d\mathscr{E}/dK$ goes to zero for all directions of propagation from the point $\mathbf{K} = \mathbf{K}_0$ at which an extremum occurs. This is a more stringent requirement which is considerably influenced by crystal symmetry, and without a careful study of this symmetry it is not possible to extrapolate from the one-dimensional problem for the position of the minima or maxima. Apart from this difficulty, the one-dimensional problem does give a useful guide showing how it is merely the wave motion combined with the regular periodic nature of the crystal structure that leads to permitted bands of propagation of energy.

In a later section the number of permitted quantum states in a band will be calculated. In a semiconductor it is expected that there are just sufficient electrons available to fill the states in the energy bands up to the valence band. Conduction electrons will then occupy the lowest energies in the next band, the conduction band. The conduction of electricity will thus be governed by the motion of wavepackets of electrons around the minima of the conduction band. It is to this problem that we next turn.

6.3 The conduction-band energy valleys

The simplest form for the minimum energy region in the conduction band is a valley, occurring at $K = 0$, $\mathscr{E} = \mathscr{E}_c$, such that for any direction of motion $d\mathscr{E}/dK = 0$; then from a Taylor expansion

$$\mathscr{E} - \mathscr{E}_c = \tfrac{1}{2}(d^2\mathscr{E}/dK^2)K^2 + \ldots \qquad 6.15$$

From comparison with eqn 6.3, it is apparent that this is identical to the behaviour of a free electron wave provided that one defines an effective mass m^* such that

$$m^* = \hbar^2/(d^2\mathscr{E}/dK^2) \qquad 6.16$$

One will then expect wavepackets centred around the minimum to behave like free electron wavepackets except for a change in apparent mass. This is the concept of effective mass in its simplest form. If an electric field is applied to a crystal then the conduction electrons are accelerated just as a free electron, but with a modified mass. In the perfectly periodic crystal as considered so far, the electron is quite free to move and could, in principle, gain a large additional energy from any applied electric field. In reality the electron loses energy by a number of collision processes, to be discussed later, so that its energy is frequently limited to a narrow range above $\mathscr{E}_c$. Without any fields applied, the electrons are naturally within a few kT of the minimum energy $\mathscr{E}_c$, T being the equilibrium temperature. Thus it is not often that there is any need for more terms in the Taylor expansion than are given in eqn 6.15.

Real conduction bands are more complicated, and indeed very difficult to calculate with any certainty.[6.3,6.13,8.18] Figure 6.4 shows roughly the form, for Si, Ge, and GaAs, of the important parts of the band structure for

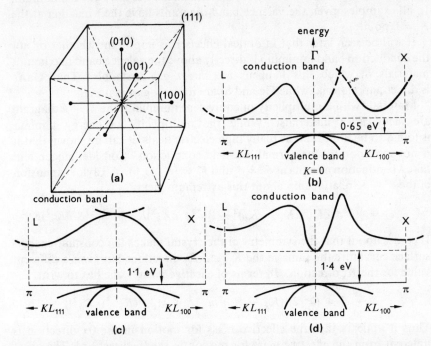

Fig. 6.4 Schematic band diagrams. (a) In Si, Ge and GaAs, the four tetrahedral bonds lie in the directions of the four diagonals of a cube. The normals of this cube define the principle axes of the crystal. There are 6 equivalent (100) and 8 equivalent (111) directions from the symmetry; this implies 8 L valleys and 6 X valleys but a single Γ valley. (b) Ge. (c) Si. (d) GaAs.

propagation along two particular directions called the (100) and (111) directions of the crystal. One notices that there are separate minima along these different directions as well as a central minimum. In all three materials shown, there is a minimum at some point $(K_0, 0, 0)$ referred to as the X minimum. There are in fact six such extrema because the symmetry of the crystal gives the same Brillouin diagram for propagation along any of the six directions $(\pm 1, 0, 0)$, $(0, \pm 1, 0)$, $(0, 0, \pm 1)$, i.e. the six principle axes of the crystal. Similarly there are 8 other minima, referred to as the L minima, situated at points $(\pm K_1, \pm K_1, \pm K_1)$ in the Brillouin zone; symmetry again ensuring the identical behaviour along these directions. Any central minimum in this structure is called Γ minimum (the nomenclature X, L, Γ being taken from group theory for the symmetry of the crystal). The precise value of K_1 and K_0 for these valleys need not concern us here; it will suffice to know that the minima lie close to the edge of the Brillouin zone. It will be important for future work to know which minima are the lowest. Thus in Ge it is the L minima, in Si it is the X minima and in GaAs it is the Γ minimum. In all examples given the valence band diagrams have their maxima at the $K = 0$ point.

It will be seen later that in optical effects it is important whether or not the conduction band minimum is directly above the valence band maximum; materials in which this happens are *direct gap* materials. Thus GaAs is direct-gap material while Ge and Si are indirect gap-materials.

Materials with a multiplicity of equivalent minima in the $\mathscr{E} - k$ diagram are *many-valley* materials. While the multiplicity of energy minima is found to be important at really high electric fields, it can be shown that it is not so important for normal low-field conduction. Consider silicon and take a conduction minima at $\mathscr{E} = \mathscr{E}_c$ and $K = (K_0, 0, 0)$. A Taylor expansion of the $\mathscr{E} - K$ relationship about this extremum yields

$$\mathscr{E} - \mathscr{E}_c = \tfrac{1}{2}(\partial^2\mathscr{E}/\partial K_x{}^2)(K_x - K_0)^2 + \tfrac{1}{2}(\partial^2\mathscr{E}/\partial K_y{}^2)K_y{}^2 + \tfrac{1}{2}(\partial^2\mathscr{E}/\partial K_z{}^2)K_z{}^2$$

It comes about that the symmetry of the crystal makes the constant-energy-surface curvature the same in the K_y and K_z directions but gives a different value for the K_x direction. In terms of effective masses one has to write

$$\mathscr{E} - \mathscr{E}_c = (\hbar^2/2)\{(K_x - K_0)^2 m_\parallel{}^{-1} + m_\perp{}^{-1}(K_y{}^2 + K_z{}^2)\} \qquad 6.17$$

Thus it appears that the effective mass for motion in the $0x$ direction is different from the effective mass for motion in the $0y$ direction! The offset in the value of K_x is another disturbing feature, though one that is readily tackled. For example in the $0x$ direction the group velocity (from eqn 6.17) is $\hbar(K_x - K_0)/m_\parallel$ so that the particle velocity is given by the change of momentum $\hbar(K_x - K_0)$. The displacement of the minimum from a central situation is then of no real importance in this context of motion. It is also found that the different effective masses for the different directions of motion in the valleys are not in fact observed for straightforward conduction processes. One must remember that for Si there are six valleys, at the lowest energy above the valence band; the six lie in pairs along three axes oriented at right angles—the principle axes of the crystal. In equilibrium, each valley contains the same number of electrons, so for conduction say in the $0x$ direction there are two valleys which give the electrons an effective mass $m_\parallel$ but four valleys, at right angles to the first two, which must give an effective mass $m_\perp$. For an applied field E the magnitude of dp/dt is the same in each valley, consequently the average rate of change of velocity is given by

$$\mathrm{d}\bar{v}/\mathrm{d}t = -eE\left(\frac{1}{3m_\parallel} + \frac{2}{3m_\perp}\right) \qquad 6.18$$

It should now be noted that the relative proportions of electrons with the different effective masses is the same even if the $0y$ or $0z$ directions were chosen. Thus eqn 6.18 turns out to be independent of the direction of the applied field and one has an effective mass for conduction given by $m^* = 3m_\parallel m_\perp/(2m_\parallel + m_\perp)$. A similar simplification holds with the eight valleys in Ge. Here too there are different effective masses which average out in their effects to give a conduction effective mass that is independent of the direction of the applied field. This simplification does not necessarily hold in the presence of a magnetic field (see Problem 6.4) but for most of the work in this book there is no need to consider these additional effects.

6.4   Holes

From Fig. 6.4, it can be seen that, for the three materials considered, the valence band has an $\mathscr{E} - K$ diagram with a maximum at the $K = 0$ point in the Brillouin zone. The detailed calculations give rise to two valence-band curves through, or very close to, the same energy at $K = 0$. We shall find later that the shallower curve has more electron states and so contains more holes. The expansion about the maximum energy can be performed by a Taylor series as in eqn 6.15 but now the curvature is negative, so that the effective mass, given by eqn 6.16, is negative! If one could obtain a few electrons at the top of a valence band which was otherwise empty, then very interesting consequences would arise with the current flowing in a sense that aided the voltage rather than opposing it as in the conventional manner. Unfortunately, the valence band is usually full, perhaps with a few electrons absent from the top. Suppose that N electrons/unit volume are needed to fill the valence band but H/unit volume of these are missing. The net momentum of the electrons could be written as $\sum_N p - \sum_H p$ where the sum has been split into the full band and the absent electrons. Now for the full band there are always as many states with positive K-vectors as with negative K-vectors of the same magnitude; the total momentum is thus always zero, so the summation over the full N-states can be omitted. The force $\mathbf{F}$ of an electric field on the electrons in a full band similarly has no net effect. Although $\mathbf{F} = \mathrm{d}\mathbf{p}/\mathrm{d}t$ for all the electrons, this force merely rearranges them and there is no net motion. The net effective force on the band then corresponds to the force on the H absent states or H missing electrons. This force must be $[-(-eH)E]$ or $+eHE$. The momentum p of each wavepacket is given by

$-m_h^*v_g$ and for the wavepackets at the top of the valence band $-m_h^* = h^2/(d^2\mathscr{E}/dK^2)$ as in eqn 6.16. Thus equating the net force with the rate of change of momentum yields

$$HeE = \mathrm{d}/\mathrm{d}t\left(-\sum_H p\right) = \mathrm{d}/\mathrm{d}t(+Hm_h^*v_g) \quad \text{or} \quad m_h^*\mathrm{d}v_g/\mathrm{d}t = +eE \quad 6.19$$

This manipulation shows that the absent state appears to behave like a positively charged particle with a positive effective mass m_h^*. This then is a simplified account of the band theory of holes which must complement the classical account on page 20. Note that an absent electron is not quite the same physical entity as a hole. The electron may be absent from a state at $\mathbf{K}_e$ but the hole has a K-vector $-\mathbf{K}_e$. The electron at the top of the valence band has a mass $m_e^* = -m_h^*$. A hole should be thought of as a wavepacket which can have momentum and energy and can collide and exchange energy with other holes and electrons. To think of a hole as merely an absent electron can lead to difficulties of thought, particularly when it comes to collisions.

Excitons[6.6]

The theory given for holes and electrons assumes a self-consistent field: the charge gives the potential, which in turn determines the motion and hence the charge distribution. The periodic potential has to be calculated in some self-consistent manner, and indeed this point has not been made in the previous work for the simple reason we have never really worried about the precise nature of the potential. In this self-consistent method, the absence of electrons in the valence band is not normally assumed, but it should be clear that a hole and an electron can attract one another, and one might expect an interaction similar to that for a hydrogen-like atom where a negative charge moves around a positive charge. This situation can occur in a solid, and such a combination of hole and electron is an *exciton*.

The simplest method for dealing with hole–electron interaction is to calculate the motion as for the hydrogen atom, except one cannot now assume that the positive charge is substantially fixed because of its heavier mass. The hole and electron have comparable masses, so that the motion must be calculated with respect to the centre of mass. If the electron, of mass m_e, is at a distance r_e and the hole, mass m_h, at a distance r_h then $m_e r_e = m_h r_h = m_e m_h/(m_e + m_h)r$ where $r = r_e + r_h$. The force between the two particles is still of course a function of r, so that the dynamics of the motion of either particle can be calculated as for the hydrogen atom, but with a reduced mass $m_{exc} = m_e m_h/(m_e + m_h)$. The calculation then follows that for the impurity state, where the electron wave function is assumed to cover many atoms and thus the electrical attraction is reduced by the dielectric constant (evaluated at optical frequencies). The lowest energy state, below the free-electron and free-hole state, is then given from either the Bohr theory or the

Schrodinger theory by $\mathscr{E}_{exc} = (m_{exc}/m)(13 \cdot 6/\varepsilon_r^2)$ eV. This is interpreted as a bound-electron state just below the conduction-band minimum, in the forbidden gap (Fig. 6.5). There is no current flow because the electron and hole move together. The free state must be an unbound electron in the conduction band and a free hole in the valence band; in between the lowest energy state and the unbound state there will be a range of states as for the hydrogen atom, though the proximity of their spacing in energy will make it difficult to observe the separate levels except at low temperatures.

Not all excitons (hole–electron states) are of this form. Some can be fixed and attached to impurity atoms, but the different forms and details must

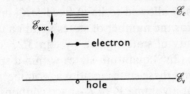

Fig. 6.5   Schematic diagram of exciton states.

be left for further reading.[6.6,9.1] Because current does not flow, excitons do not enter directly into conduction theory. They will be found to be more important in the discussions on the interaction of light with solids (Chapter 9).

6.5   Quantum states

In Chapter 3 statistical techniques were applied to find the number of electrons and holes occupying the appropriate bands. This calculation was not completed in detail because it required a knowledge of the number of quantum states around a given energy. The knowledge gained in this chapter about waves in crystals enables such a calculation to be made. For simplicity a cubic crystal structure will be assumed, so that N^3 unit cells form a solid cube of sides A with $A = NL$, L being the unit cell length. To avoid effects caused by the boundaries, it will be assumed further that there is an ensemble of cubes stacked together, and that the only permitted wave functions are those which make each cube of this ensemble identical to any other cube. This means that the wavelength can only be integral parts of the cube sides, for propagation along the direction of one of the cube edges. Thus, take K_x, with $0x$ along one of the cube edges

$$K_x = 0, \pm \frac{2\pi}{A}, \pm \frac{4\pi}{A}, \pm \frac{2n\pi}{A}, \ldots \text{ to } \pm \frac{\pi}{L}$$

The outer limits to K_x are given from the condition that K_x lies within the Brillouin zone as discussed in section 6.2; this limits $2n$ to be less than N

for any given band. Now for each value of K_x there are N values of K_y and N values of K_z giving N^3 permissible states for the K vectors. Allowing for the two possible spin orientations, one has $2N^3$ electron states within one energy band. One notices that this result immediately shows that an insulator, where the valence band is full, has to have an even number of valence electrons per unit cell. The converse is not always true because, in three dimensions, the energy of the valence band can be higher in energy in one part of the Brillouin zone than the conduction band in some other part of the zone.

It is important to note that the states lie *along* the $\mathscr{E} - \mathbf{K}$ surface and do not *fill* the valley as some text-books show. Figure 6.6 shows the effect of the numbers of quantum states in a unit energy range $\delta\mathscr{E}$, given that the permitted states are only allowed to lie at fixed $|\delta K|$ intervals. The task is to find an expression for the number of states within a unit energy range so that the effective density of states, given by eqn 3.22, can be evaluated in detail. Now there are $(2N^3)$ quantum states within a spatial volume $(NL)^3$ for the cubic crystal, and these states require a 'volume' in K-space given by $(2\pi/L)^3$. It follows that a volume V_k of K-space contains

$$V_k[2N^3L^3/(NL)^3 \cdot (2\pi)^3] = V_k/4\pi^3$$

quantum states per unit volume of real space. Now from eqns 6.15 and 6.16, a change in energy from $\mathscr{E}$ to $\mathscr{E} + \delta\mathscr{E}$ occupies a spherical shell in K-space such that the shell has a 'radius' $[2m^*(\mathscr{E} - \mathscr{E}_c)/\hbar^2]^{1/2}$ and 'thickness' $\frac{1}{2}[2m^*/(\mathscr{E} - \mathscr{E}_c)\hbar^2]^{1/2}\,\delta\mathscr{E}$. Its 'volume' in K-space is thus $2\pi(2m^*/\hbar^2)^{3/2}(\mathscr{E} - \mathscr{E}_c)^{1/2}\,\delta\mathscr{E}$. Hence the number of states between $\mathscr{E}$ and $\mathscr{E} + \delta\mathscr{E}$ is given by

$$N(\mathscr{E})\,\delta\mathscr{E} = 4\pi(2m^*/h^2)^{3/2}(\mathscr{E} - \mathscr{E}_c)^{1/2}\,\delta\mathscr{E}/\text{unit volume of crystal} \qquad 6.18$$

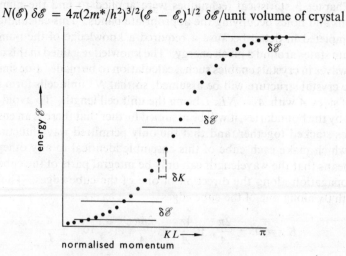

Fig. 6.6 Variation of density of electron states.

If one bears in mind that the electrons will only have a significant probability of lying at the lowest energies it will be apparent that the structure of the $\mathscr{E} - K$ diagram at a few kT above the minimum will be quite unimportant. One may then integrate from $(\mathscr{E} - \mathscr{E}_c) = 0$ to ∞ in eqn 3.21, using a standard integral

$$\int_0^\infty u^{1/2} \exp(-\beta u)\, du = \tfrac{1}{2}(\pi)^{1/2}\beta^{-3/2}$$

so that in eqn 3.21 $N_c = 2(2\pi m^* kT/h^2)^{3/2}/\text{unit vol.}$ 6.19

This, it must be remembered, is for one 'valley' in the conduction band; in many-valley semiconductors such as Si, one must multiply N_c by M, the number of such valleys, before it can be used in eqn 3.21. For material like Si, where the effective mass varies with direction, the constant-energy shell is an ellipsoid rather than a sphere. The effective mass for the density of states function N_c then is given by $m^* = (m_1 m_2 m_3)^{1/3}$ where m_1 etc. are the three principal effective masses. In Si, m^* for the density of states would be $(m_\parallel m_\perp^2)^{1/3}$.

Similar results follow immediately for holes, and are not repeated here in detail. The two valence bands with different curvatures give different effective masses for the holes—light holes m_1 and heavy holes m_h

$$N_v = 2(2\pi kT/h^2)^{3/2}[m_1^{3/2} + m_h^{3/2}]/\text{unit vol.}$$ 6.20

It can be seen that the heavy holes will contribute most significantly.

6.6   Emission over a potential barrier

An important effect in physical electronics is the rate at which electrons with sufficient energy can cross a barrier of energy $\mathscr{E}_b$. The analysis is of importance in understanding emission of electrons into a vacuum from a hot metal[6.7] or semiconductor, and in understanding current flow from a metal to semiconductor or vice-versa.[6.8]

Consider a potential barrier across the $0x$ direction such that only electrons with energy greater than $\mathscr{E}_b$ can cross. If the barrier height is well above the Fermi level, then Boltzmann statistics will apply to the relevant electrons. The discussion will be limited to a 'central' valley with a uniform effective mass. Only those electrons with kinetic energy in the $0x$ direction greater than $\mathscr{E}_b$ will in fact be able to surmount the barrier; transverse energy will not help in a true one-dimensional problem. Thus the range of K_x is limited to values greater than K_m given by $K_m^2 = 2m^*(\mathscr{E}_b - \mathscr{E}_c)/\hbar^2$ (using eqns 6.16 and 6.15). In this discussion it will be convenient to use the elemental volume of K-space given by $dK_x \cdot dK_y \cdot dK_z$. The current density of the electrons crossing the barrier is given from their velocity, $(\hbar K_x/m^*)$, together with the probable number density in each allowed range of K-vectors, thus:

$$J_x = e \exp -(\mathscr{E}_c - \mathscr{E}_f)/kT \iiint (\hbar K_x/m^*) \times$$
$$\exp\left[-(\mathscr{E} - \mathscr{E}_c)/kT\right](dK_x\, dK_y\, dK_z/4\pi^3)$$

Rearranging and using eqn 6.16,

$$J_x = e \exp -(\mathscr{E}_c - \mathscr{E}_f)/kT \int_{K_m}^{\infty} \frac{\hbar K_x\, dK_x}{4\pi^3} \times$$
$$\int_{-\infty}^{+\infty} dK_y \int_{-\infty}^{+\infty} dK_z \exp\left[-\left(\frac{\hbar^2}{2m^*kT}\right)(K_x{}^2 + K_y{}^2 + K_z{}^2)\right]$$

From standard integrals $\left[\int_{-\infty}^{+\infty} \exp -ax^2\, dx = (\pi/a)^{1/2}\right]$

$$J = (m^*/m)AT^2 \exp -(\mathscr{E}_b - \mathscr{E}_f)/kT \qquad\qquad 6.23$$

where $A = (4\pi emk^2/h^3) = 1{\cdot}2 \times 10^6\, A/m^2 K^2$ is called the 'Richardson constant' and $(m^*/m)A$ the 'modified Richardson constant'.

It may be noted that the result of eqn 6.23 is independent of $\mathscr{E}_c$. However, for electrons well above the Fermi level in a metal the relationship of eqn 6.15 still holds, but with $\mathscr{E}_f$ replacing $\mathscr{E}_c$. Thus the above analysis also gives directly the emission of electrons from a metal into a vacuum. The emission current density is $J = AT^2 \exp -e\phi/kT$ where ϕ is the work function $(e\phi = \mathscr{E}_b - \mathscr{E}_f)$.

6.7 Mobility and phonons

The concluding section of this chapter considers the major causes of resistance to the motion of electrons in a semiconductor. It was seen in Chapter 3 that the electrons were only free to gain momentum from an applied electric field for a mean time τ_m, and yet in the last few sections it has been demonstrated that the electrons are free to move continuously through a periodic crystal. Clearly these conflicting views need reconciliation. It has already been hinted that the reconciliation comes from the fact that the real crystal is not perfectly periodic in structure. With this in mind we examine some of the basic collision processes.

(a) Electron–electron collisions

Electrons collide among themselves as they move in a crystal. The periodic potential assumed in the simple analysis of the wave motion is an average potential produced by the atoms and their surrounding electrons. This is used to calculate the electrons' motion in a self-consistent manner. The fact that real electron wavepackets are localized implies fluctuations from the perfectly periodic potential, so that electron wavepackets will scatter one another. In particle terms, one says that the electrons collide with each other. Both energy and momentum are conserved in these interactions or collisions, so that the electron gas as a whole could continuously gain momentum and energy if these were the only collisions to be considered. Electron–electron collisions thus only redistribute the energy among the electrons and are not directly responsible for resistance to the electrons' motion in the lattice under an applied electric field.

(b) Defects

Defects in the crystal structure are caused by local changes in the binding pattern between atoms. They destroy the perfection of the periodic structure of the atoms in the crystal. There is an increased reflection of any electron wave incident on the defect and the 'collision-free' time is reduced. A common defect is a dislocation where a row of atoms have slipped along a plane. This has some effects which are analogous to a free surface: defect states, changes of local potential and enhanced recombination. Material with a high density of defects will then have a shorter free time for its electronic motion and a lower mobility (as well as other undesirable features). Modern technology has helped to produce very high-quality material with low concentrations of defects, so in good material this is not a fundamental source of electrical resistance.

(c) Impurities

A third source of disruption to the periodic lattice is the addition of impurities. There are two types: the charged impurity and the uncharged. The latter are often unwanted and a great amount of effort and money has helped to produce pure materials, so once again these sources of collision are not fundamental to the electrical resistance.

The charged impurities of the donors and acceptors which are ionized are fundamental, because without these impurities the semiconductor cannot conduct or control the flow of charge. An important concept in collision theory is the collision cross-section;† this cross-section for the electron-charged impurity collision decreases as $(1/v)^4$ where v is the average electron velocity. In mechanistic terms one is saying that the faster an electron moves then the less it is deflected by a local change of potential. If σ_i is the collision cross-section for this process then there are $N_i \sigma_i v$ collisions per

---

† The collision cross-section is an *effective* area σ assigned to an obstacle on the assumption that the incident particle has zero volume. An incident particle will then have $N\sigma$ collisions per unit distance of travel, where N is the density of obstacles. If v is the average velocity of the particle there are $N\sigma v$ collisions per unit time.

unit time where N_i is the impurity density. The reciprocal of this collision rate will give the mobility through eqn 3.4, so that the mobility will be expected to vary as v^3/N_i. Now the electrons velocity at room temperature is around 10^5 m/s average value. In most semiconductors this mean velocity will change but little with electric fields until the fields are very strong. Thus the kinetic energy is determined primarily, at low electric fields, by the lattice temperature T; the impurity mobility varies as $[T^{3/2}/N_i]$. A detailed calculation, first given by Conwell and Weisskopf,[6.10] gives an impurity mobility μ_i from[6.12]

$$\mu_i \simeq \frac{0.54(\varepsilon_r/12)^2(T/300)^{3/2}(m/m^*)^{1/2}(10^{24}/N_i)}{1 + 0.24 \log_e[0.02 + 0.98(T/300)^2(\varepsilon_r/12)^2(10^{24}/N_i)^{2/3}]} \; \text{V} \cdot \text{s/m}^2$$

It must be emphasized that this mobility includes only the impurities and no other effects. Practical semiconductors usually have mobilities below unity in SI units, so that the impurity scattering is not the main limit on the mobility until impurity densities are rather high, or temperatures low.

(d) Lattice vibrations—phonon–electron scattering

The atoms in a solid vibrate with thermal energy about their ideal positions in the periodic lattice. This vibration perturbs the perfection of the periodic structure and scatters the electrons. As for impurities and defects, the exchange of energy is between the electron and the crystal, so that the electrons do lose energy and momentum and the electrical resistance is increased, or equally the electron mobility is decreased.

Because the atoms in a solid are bound together, the vibration of any one atom is propagated to the other atoms by waves—lattice vibrational waves. The particles associated with these waves are called *phonons*, and one refers to phonon–electron collisions to describe the scattering process. The physical outline of the interaction is as follows. Thermal energy gives rise to lattice waves which move through the crystal and locally deform it, compressing it in some places and expanding it in others. Provided that the deformation occurs over several atoms, the electron wave will behave as if the periodicity of the crystal has slightly changed. If the simple model of eqn 6.14 is taken as a guide, it can be seen that the change of periodicity implies a change of $\mathcal{E}_L$ and so a change of the band edges or energy extrema. The potential energy of the electron wavepacket will thus change, its K-vector will slightly alter and the wave will be slightly reflected. In particle terms one says that a phonon has collided with an electron or vice-versa. The type of scattering just considered is known as deformation-potential scattering, the deformation potential being the change in band-edge potential per unit dilation of the crystal (dilation is the fractional change of volume $\delta V/V$).

Phonons

Phonons in a crystal have permitted ranges of energies arising, as for electron waves, because of the periodicity of the crystal. The frequency associated with the movement of phonons is a measurable frequency and not just a convenience, as it is for electron waves. The phonon spectrum ranges from zero to typically 10^{13} Hz or thereabouts. The Brillouin diagram for phonon waves can be calculated from a knowledge of the crystal symmetry and structure, though realistic calculations are very difficult. There are also clever techniques which consist in principle of making diffraction experiments with neutron quantum waves, analogous to electron diffraction for finding crystal structure. Figure 6.7 shows a typical result for a $\omega - k$ diagram for

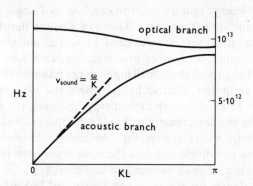

Fig. 6.7 Schematic diagram of phonon waves (in crystal with two atoms/unit cell).

these lattice waves or phonon waves. The lowest frequencies correspond to sound waves in the crystal and this gives the low branch the name of the *acoustic* branch, though the frequencies range well above the normal idea of acoustic frequencies. Indeed the excitation of acoustic waves at microwave frequencies has proved to be a very fascinating new field of research with several practical devices, especially for processing signals required for radar and communications.[2.22] However, these devices tend to use *surface waves*, waves with their energy confined to the surface of the crystals, and not the bulk waves that are important in the present discussion. The reader must not forget that the propagation of such high-frequency waves will usually be accompanied by considerable attenuation, or loss, as the waves travel through the crystal (there are some exceptional materials with very low acoustic wave losses, even at frequencies above 10^9 Hz). As for electron waves, the periodicity of the crystal forces the propagation to stop whenever the repetition length of the crystal pattern is a half-wavelength (see Problem 6.3). Hence $d\omega/dk = 0$ when $kL = \pi$, where L is the repetition length for the unit cell. The higher branch of the lattice waves in Fig. 6.7 is called the *optical* branch and in principle there could be further branches. This optical branch arises typically in a crystal where there are just two atoms per unit cell and the two atoms vibrate about their centre of mass. Such vibrations are at low 'optical' frequencies and are almost independent of their neighbouring atoms. It is this independence which gives the low group velocity or almost constant frequency. Although only one dispersion diagram has been shown, there are modifications which depend on whether the atoms are vibrating transversely or longitudinally in relation to the propagation direction. The reason why the associated phonons are called transverse or longitudinal phonons is obvious.

The concept of phonons will allow several simple deductions to be made provided that the collision process is considered in terms of colliding particles. The maximum energy that an electron can lose to a single phonon is $h\omega_{opt}$, the energy of an 'optical' phonon. One can also see that large changes of momentum (or K-vector) are required to scatter an electron from one conduction minimum to another in a material like silicon. These changes can be accomplished with the help of high-energy phonons but not low-energy phonons. Thus at low electric fields, where the electrons do not have enough energy to excite many high-energy phonons, the scattering of the electrons will be confined mainly within the compass of a single conduction-band valley (intra-valley scattering). At higher electric fields the inter-valley scattering can dominate, with the optical phonons playing an important role in the exchange of momentum and energy.

The formal theory for phonons can be carried but little further in this elementary text. As phonons do not have any spin, it is expected (see footnote on page 10) that they will not obey the Pauli-exclusion principle and in this respect they are similar to photons and obey Bose-Einstein statistics (see Chapter 9). Electron–phonon collisions are frequently the fundamental processes determining the mobility of the charge carriers. The detailed theory again requires more quantum mechanics than is assumed here, and so this is left for further reading. Roughly speaking, the density of the phonons which actively collide with the electrons increases as the temperature so that, with the average velocity increasing as $T^{1/2}$, the low field mobility for the electrons in a crystal is expected to vary as $T^{-3/2}$. However, practical results show considerable deviations from this simple theory. Figure 6.8 indicates qualitatively the changes of mobility that might be expected in a crystal as the temperature changes. At high temperatures the phonon–electron collisions will dominate, while at low temperatures the impurity scattering becomes more important, provided that the impurities are still ionized to give free carriers. At high electric fields new effects occur, but these must be left for Chapter 8 and further reading. It will still be found that the concept of phonons is helpful in appreciating the physics.

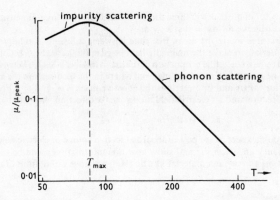

Fig. 6.8 Schematic diagram of variation of mobility with temperature. T_{max} may be lower or higher depending on material. Impurities may not ionize if temperature is too low, and material shows high resistivity even though mobility may be high theoretically.

PROBLEMS

6.1 A transition of an electron from one quantum state to another quantum state at 100 eV lower in energy takes 10^{-12} s in some particular atom, and a soft X-ray photon is emitted. Show that the uncertainty of the photon's energy is the order of 0·005 per cent.

An electron is confined to a unit cell of dimensions 0·5 nm cube. Show that the uncertainty in the momentum is greater than 10 times the average thermal momentum, at room temperature.

6.2 An electron wave with kinetic energy E_k is incident upon a potential barrier that is L thick and E_b in height. The potential energy remains constant throughout the system and so does the mass of the electron. Show that if $E_k \ll E_b$ the number of escaping particles is approximately $(16E_k/E_b) \exp\left[-(8mE_b/h^2)^{1/2}L\right]$ fewer than the incident number of particles on the barrier. Taking $E_k = 25$ meV (a thermal energy at room temperature), $E_b = 1$ eV, $L = 10$ nm, and $10^{25}/\text{m}^3$ electrons on one side of the barrier, estimate the number of carriers on the other side of the barrier. Using an average thermal velocity, estimate the maximum current density of emission. (*Hint:* the barrier is assumed thick enough for the two edges to be treated independently because of the decaying amplitudes of the waves. At each face, the incident, reflected and transmitted waves are continuous in amplitude, i.e. $\psi_i + \psi_r = \psi_t$ and also from Schrodinger's equation it follows that $\psi_i' + \psi_r' = \psi_t'$ where $\psi' = \partial\psi/\partial x$).

6.3 In the one-dimensional wave motion in a periodic structure, $d\mathscr{E}/dk = 0$ whenever the period a of the structure is an integral number of half-wavelengths apart (i.e. $ka = \pi$). This follows from constructive interference of the waves preventing propagation of energy. The following model illustrates this point.

Consider identical obstacles placed $n\lambda/2$ apart with n integer. Show that the combined reflection coefficient of a pair of obstacles is given by $R = r + T$ where T is the combined transmission coefficient given by $T = t^2/(1 - r^2)$. The parameters r and t are the reflection and transmission coefficients of the isolated obstacles. Hence combine 2^m such obstacles showing that every time m increases by one unit $R_{m+1} = R_m + T_{m+1}$ and $T_{m+1} = T_m^2/(1 - R_m^2)$. In each case the reflection and transmission coefficients are referred in position to the front obstacle (i.e. as if the front obstacle were the only effective obstacle in position).

Hence show that, if $r = 0$ and m is very large, there are constants D and C such that

$$R_m \sim 1 - C(\tfrac{1}{2})^{m/2} \qquad T \sim D(\tfrac{1}{2})^{m/2}$$

Hence show that propagation of energy is not possible and so one must expect $d\mathscr{E}/dk = 0$.

If $r = 0$ then clearly the obstacles are transparent, and so $d\mathscr{E}/dk$ does not go to zero. However, in the *reduced Brillouin zone* scheme the *average* value of $d\mathscr{E}/dk$ goes to zero and this is the more general result for any set of obstacles in three dimensions. Thus in a three-dimensional diagram one does not always observe $d\mathscr{E}/dk = 0$ at $k = \pi a$.

6.4 At low temperatures where the mean free time between collisions is relatively long, it is possible in some materials to make the magnetic field large enough for the orbit of the electron in the magnetic field to be a negligible proportion of the mean free-path-length between collisions. Collisions can then be ignored. Consider a material where in the $0x$ direction the effective mass is m_x and similarly for the $0y$ and $0z$ directions the effective mass is m_y and m_z. For a magnetic field along the $0z$ direction and an electric field along the $0x$ direction show that

$$\ddot{p}_x + [(eB_z)^2/m_x m_y]p_x = e\dot{E}_x$$

Explain why one would expect to see power absorbed from a source of electric field alternating at an angular frequency $eB_z/(m_x m_y)^{1/2}$ and show how this information could be used to obtain the effective masses $m_{\parallel}$ and $m_{\perp}$ in a material like Si. (Hint: change orientation of field)

General references

    KITTEL, C. Reference 2.1.
6.1 SMITH, R. A. *Semiconductors*. Cambridge University Press, 1959.
6.2 SHOCKLEY, W. *Electrons and Holes in Semiconductors*. Van Nostrand, 1950.

Special references (* marks an advanced level)

* Pseudo-potentials

6.3 HEINE, V. The pseudo-potential concept. *Solid State Physics* (Academic Press), **24** (1970), 1–36.
    The rest of the volume also contains more advanced work.
6.4 HARRISON, W. A. *Solid State Theory*. McGraw-Hill, 1970.

Floquet's theorem

6.5 INCE, E. L. *Ordinary Differential Equations*. Dover, 1956.

* Excitons

6.6 KNOX, R. S. Theory of Excitons. *Advances in Solid State Physics*. Academic Press, 1963.

Thermionic emission

6.7 BECK, A. H. W. and AHMED, H. Reference 1.1; Chapter 6.

Emission into semiconductors

6.8 SZE, S. M. Reference 1.5, Chapter 8.
6.9 CROWELL, C. R. The Richardson constant for thermionic emission in Schottky barrier diodes. Solid State Electronics, **8** (1965), 395.

Collisions

6.10* CONWELL, E. and WEISSKOPF. Theory of Impurity scattering in semiconductors. *Phys. Rev.* **77** (1950), 388–390.

6.11 ZIMAN, J. M. *Principles of the Theory of Solids.* Cambridge University Press, 1964.
6.12 MOLL, J. L. *Physics of Semiconductors.* McGraw-Hill, 1964.

Band diagrams*

HERMAN, F. The electronic energy band structure of silicon and germanium. *Proc. IRE,* **43** (1955), 1703–1735.

7

Devices using Metal, Semiconductor, Insulator Combinations

7.1 The Schottky barrier

The bipolar transistor is a relatively complex device to manufacture in its modern form. It has several photolithographic steps—etchings and diffusions, together with its metalization. Each manufacturing step potentially lowers the yield of devices that work in a given batch. In this chapter it will be seen that there are other devices that can be made with a smaller number of processing steps by using metals and insulators in conjunction with the semiconductor material. Indeed these devices will have additional merits of their own besides simplicity of construction. The opening discussions are about one of the devices most 'simple' in construction. It consists of a metal layer on top of an n-type semiconductor and is known as a *Schottky Barrier* diode.

Figure 7.1a gives band diagrams for a metal (a) and semiconductor (b) and shows the relationship of the vacuum level to the Fermi level. In the metal, the vacuum level is at an energy $e\phi_m$ above the Fermi level and ϕ_m is the *work function* (measured in electron volts). Only those electrons which can surmount such a potential barrier can escape from the metal into the vacuum. In the semiconductor the analogous parameter is the *electron affinity* χ, except that this is measured from the bottom of the conduction band. Depending on the differences between ϕ_m and χ, one finds different current–voltage relationships for these metal–semiconductor junctions. In the ideal situation the metal and semiconductor are in close metallurgical contact, with an abrupt change. However, there cannot be infinite electric fields in the vacuum close to the materials, so that there cannot be an *abrupt* change of potential in the vacuum level at any junction. The vacuum levels then align at the junction, though they may change away from the junction. Indeed the band potential must change in the semiconductor so that the Fermi level in the bulk of the semiconductor can be aligned with that in the metal. The potential changes are brought about by a redistribution of the charge in the semiconductor, there being no corresponding change of potential inside the metal because the high conductivity maintains electrical neutrality. With these considerations in mind one can sketch the band diagrams for a number of interesting examples.

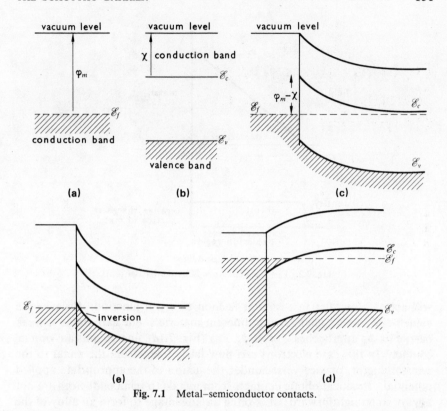

Fig. 7.1 Metal–semiconductor contacts.

Figure 7.1c shows a junction of an n-type semiconductor with a metal such that $\phi_m > \chi$. Electrons being at a higher energy will at first diffuse out of the semiconductor, leaving behind a depletion region that will make the semiconductor positive with respect to the metal and so prevent further diffusion. The Fermi levels are then aligned. This is the situation normally envisaged for the Schottky barrier diode. A dense but thin layer of electronic charge will accumulate at the interface of the metal and the semiconductor, while there will be a positive charge of the ionized donors in the depletion region in the semiconductor. The two charges will cancel to preserve the total electrical neutrality and the required potential change will be established by adjustment of the depletion length (see Fig. 7.2). Notice that there is always a potential barrier, $\phi_m - \chi$, to the electron flow *from the metal into the semiconductor* and that this is not altered by any applied potential. The barrier to the electron flow from the semiconductor into the metal is there only because of the diffusion process leading to a depletion. A positive potential applied to the metal will reduce this barrier and lead to electrons flowing *from the semiconductor into the metal*. Thus a Schottky barrier diode

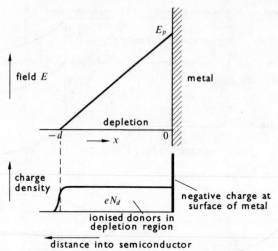

Fig. 7.2   Field and charge in Schottky barrier.

will act as a rectifier in a similar fashion to a p-n junction. Because the values of ϕ_m and χ change with different materials, one can obtain a weak barrier as ϕ_m approaches χ or, if $\phi_m < \chi$ (Fig. 7.1d), one obtains an 'ohmic' contact. In this case electrons can flow fairly freely from the metal to the semiconductor, or vice versa, under the action of the appropriate applied potential. Practical ohmic contacts from metals to semiconductors are not always so straightforward; it is often an advantage to form an alloy of the semiconductor and metal so that this forms a transition region between the two materials. The final example, in Fig. 7.1e, shows an extreme example for $\phi_m > \chi$ where the Fermi level at the junction is now close to the valence band edge. The semiconductor at this point then contains holes and is said to be *inverted*. Here it will be assumed that inversion is either absent or is negligible in its effect on the operation of the Schottky diodes.

In principle, the same features can be shown for a contact between a metal and p-type semiconductor. The reader should sketch such diagrams for himself and note that now $\phi_m < \chi$ will in general give the barrier contact while $\phi_m > \chi$ will give the ohmic contact. One must, of course, note that holes move into the semiconductor because electrons leave the valence band to enter the metal: holes do not normally come directly from the metal. Returning to the Schottky barrier (Fig. 7.1c) the mathematics of the depletion region follows that of the idealized p^+-n junction. Using the notation of Fig. 7.2, where a forward bias V_a has been applied, Gauss' theorem and the condition $E = 0$ at $x = -d$ give

$$E = (eN_d/\varepsilon_o\varepsilon_r)(d + x) = -dV/dx \qquad 7.1$$

Integration of the field with $V = 0$ at $x = 0$ then gives

$$V(x) = -(eN_d/2\varepsilon_o\varepsilon_r)(2dx + x^2) \qquad 7.2a$$

It may be seen, Fig. 7.1c, that the change of potential for the conduction band is given by $[\phi_m - \chi - (\mathcal{E}_c - \mathcal{E}_f)]$ when no bias is applied, so that for a forward bias of V_a one must have the depletion absorbing a voltage

$$V(-d) = \phi_m - \chi - V_a - (\mathcal{E}_c - \mathcal{E}_f) = (eN_d/2\varepsilon_o\varepsilon_r)d^2 \qquad 7.2b$$

The depletion width d, of course, widens on reverse bias and decreases on forward bias exactly as in a p-n junction.

Current flow

If one tried to estimate the current flow for the Schottky barrier diode in the same manner as for the p-n junction, one would calculate the electron concentration at the edge of the depletion region adjacent to the metal by the use of the Boltzmann assumption. For a forward bias voltage V_a the concentration at the edge would be increased by a factor $\exp eV_a/kT$ over the equilibrium value. However, when such electrons entered the bulk of the metal they would be far more energetic than most electrons because they would come from an energy $\phi_m - \chi$ above the Fermi level. Such electrons are called 'hot' electrons and the device is sometimes called a 'hot carrier diode'. The energy of these carriers in the metal implies that they diffuse away from the edge so rapidly that it is impossible to maintain an excess concentration of electrons at the boundary. Any calculation based on the assumption of such an excess must therefore fail. An opposite assumption supposes that thermal equilibrium concentrations are maintained at the interface as well as in the bulk of the metal. Problem 7.1 shows how the current can be calculated using this assumption. The value obtained depends rather critically on the precise potential variation that is assumed. One classical result for the Schottky barrier assumes the variation given in eqn 7.2a. However, such a theoretical result sometimes appears to give higher values of current than those observed. The discrepancy arises because the electrons can only cross a potential barrier at a certain rate (see section 6.6). This process is that of *thermionic emission* and is probably best known in relation to the emission of electrons into a vacuum from a white-hot metal or some specially prepared surface. It is a standard method of obtaining electrons in devices like cathode-ray tubes and thermionic valves. For emission from a metal into vacuum the electrons have to surmount a barrier ϕ_m, but equally one can have a similar process of emission of the energetic electrons from a metal into a semiconductor, and vice versa. With small current flows the changes from equilibrium are not too large and the distribution of energies amongst the electrons still corresponds to a Fermi-Dirac distribution. The arguments leading to the result of eqn 6.23 therefore still hold. The electrons in the semiconductor have to surmount a barrier $(\phi_m - \chi - V_a)$ above the Fermi level, while the opposing current from the metal into the semiconductor always has a barrier of $\phi_m - \chi$. The use of eqn 6.23 shows that there is a net current density for the electron flow from the semiconductor into the metal given by

$$J = A^*T^2[\exp - e(\phi_m - \chi)/kT][\exp eV_a/kT - 1] \qquad 7.3$$

Practical measurements on good Schottky barriers show current and voltage relationships which vary over many decades of current as $I = I_s \exp(eV_a/mkT)$ where m is usually a few per cent higher than unity. There are several important effects which create significant changes from the simple theory given above. These are listed in the next paragraph.

An interfacial layer between the metal and semiconductor is usually present and has uncertain electrical properties but allows fields to appear across it. This, in turn, allows significant changes to the theoretical barrier height $(\phi_m - \chi)$. On the simple theory, the barrier height varies directly with the work function of the metal, but practical measurements show that this dependence is reduced by a factor of four to ten, depending on the semiconductor. It is found that surface states and surface charges arise because of the sudden discontinuity in the electronic binding of the atoms. These surface states are discussed later; their effect here is to cause very large changes of charge at the surface of the semiconductor unless the Fermi level is very close to some zero surface-charge condition determined by surface conditions rather than the metal. The Fermi level is thus pinned close to a given level (typical barrier voltages for n-types Si and GaAs lie around 0·8 V, varying a little with the metal used for the barrier). A small effect is also caused by the fact that a charge is attracted to a metal surface by its image in the metal plane. This leads to a lowering of the potential barrier, called the *Schottky effect*, and is well known in the theory of emission from metals into vacuum.[6.7] In the Schottky diode this lowering of the barrier is around 20 mV, but changes a little with variations in forward bias. The emission theory may fail if the rate at which charges can diffuse across the layer is lower than the rate at which they can be thermally emitted. The theory therefore has to be modified for low mobility semiconductors, but this more detailed work is left for further reading.[7.1] On the constructional side, great care has to be taken to ensure good metallurgical contact between the metal and the semiconductor—but without any alloying. Alloying of the metal and semiconductor could lead to poor rectification properties.

Minority carriers

The contact of the metal with the semiconductor will maintain the electron concentration in the valence band at its thermal equilibrium value. There are thus the equilibrium number of holes present, and on forward bias these holes will contribute to a current flow exactly as in a p-n junction. The result of eqn 4.12 would hold unaltered. However, from the value of I_s, the majority carrier flow of electrons is usually *much* larger† than the corresponding flow of the minority carriers (holes). The majority carriers then

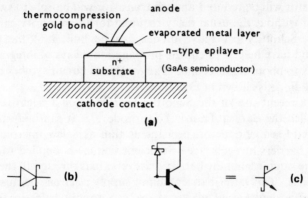

Fig. 7.3 Schottky diodes. (*a*) Possible construction. (*b*) Circuit symbol. (*c*) Schottky-protected transistor.

carry the current and the charges can be removed in a time determined principally by the transit time of these majority carriers across the depletion region. Recombination is no longer important in determining the removal of stored charge, and consequently the diffusion capacitance is negligible in a well-constructed Schottky diode. The depletion capacitance can be kept small by fabricating diodes with small cross-sectional areas. Such diodes can then be made to switch from the forward-conduction, low-impedance state into the reverse-conduction, high-impedance state in a few picoseconds, provided that the voltage changes fast enough. The principal time constant is formed by the series resistance of the undepleted semiconductor material with the depletion capacitance. Parasitic inductances associated with the package can also be important. Figure 7.3*a* indicates schematically one possible realization of a Schottky barrier diode.

An initial application of these diodes was for the detection of very high frequencies—they are good rectifiers even at 10^{10} Hz. They are now commonly used as non-linear conductances for mixing signals of different frequencies (again usually above 10^9 Hz). Recently the improved technology

† At a given forward voltage value.

has made them considered for high-power a.c. rectifiers where the lower forward-voltage drop should make them more efficient than the p-n counterpart. For Si Schottky diodes, one typically considers 0·4 V as the forward 'barrier' voltage ϕ (Fig. 1.2) as compared to 0·6 for the Si p-n junction. If such a Schottky diode is placed in parallel with the base–collector p-n junction of a transistor, then, in the saturated condition where the base–collector junction is forward biased, it will be the Schottky diode that will carry the current rather than the p-n junction (because of this lower ϕ). Thus, for a transistor so protected, the minority charge stored in the base of the transistor will be reduced and the transistor will be able to switch out of saturation without the usual delay required for the minority carriers to be removed. Schottky protective diodes can be built into the fabrication process and have helped to cut the propagation delays of integrated circuit logic units to about nanosecond. The combined circuit symbol of the transistor and diode is shown in Fig. 7.3c.

A more recent use of the Schottky barrier is for a negative resistance device called the Barrier Transit Time diode.[7.13] It can be formed from a thin layer of semiconductor, perhaps as thin as a few micrometres, with Schottky barriers at each face. Sufficient voltage is applied to *completely* deplete the sample from the barrier on reverse bias through to the barrier on forward bias. This latter barrier cannot supply *electrons* because the semiconductor is depleted of electrons—it can supply *holes* in the manner discussed above. The holes drift through the sample and at frequencies around $3/4\tau$, where τ is the transit time of the carriers, it is found that power can be given out rather than absorbed. The device shows promise on account of the low level of noise that it adds to any signal—but its mechanisms must be left for further reading. It is mentioned here to remind the reader that in suitable circumstances, 'minority' holes can sometimes be associated with Schottky barriers.

7.2   The Schottky-gate field-effect transistor

The Schottky barrier forms a depletion region exactly as in a p^+-n junction. Consequently one can make a field-effect transistor with a Schottky gate on an n-type channel which operates in the same manner as the p-gate on n-channel FET discussed earlier. Figure 7.4 shows the schematic construction of one of these devices. Of particular interest is the substrate material shown here as semi-insulating Gallium Arsenide. In section 2.7 it was mentioned that certain deep impurities could compensate the semiconductor material and make it appear almost intrinsic; oxygen or chromium, for example, are deep impurities that make GaAs become semi-intrinsic, with a very high resistivity (Problem 2.2). GaAs then forms an excellent insulating

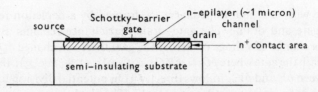

Fig. 7.4   A GaAs Schottky-gate FET (for very high frequencies).

substrate on which can be grown a thin n-type epilayer for the channel. The source and the drain can be metallic-alloyed contacts or specially grown areas of n$^+$-epilayer. The latter type of contact provides lower series resistance between the source and drain, although extra processing is required because the contact areas have to be specially etched out and then regrown in an epitaxial reactor. These types of transistors have operated at frequencies as high as 18 GHz, albeit with only a very low power. The FET is in fact usually considered as a low-power, low-noise device to be used for amplifying signals in the receiving stages of very-high-frequency communication links. The Schottky gate FET promises to be very useful in these applications.

The theory for the device at low frequencies follows that already given for the JUGFET and the same equivalent circuit can be used (see Fig. 5.18a). However, modifications must be made for high frequencies and a simplified account is presented here which emphasizes parameters that are similar to those required for the bipolar transistor. Figure 7.5a indicates that the gate-source capacitance C_{gs} is now split into two parts: a primary part forming a gate capacitance C_g to the channel and a secondary part forming the stray C_s between the gate and source. The channel resistance r_c is principally assigned between the gate and the source because it is assumed that

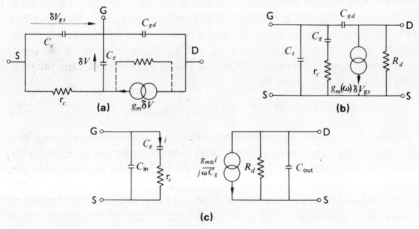

Fig. 7.5   High-frequency FET circuits. (a) Simplified view of gate capitance. (b) Rearrangement of (a). (c) Limiting circuit at high frequencies.

there will be a high impedance, R_d, formed by a depletion region between the gate and drain. The gate–drain capacitance remains as in Fig. 5.18. For a change of voltage δV across the gate there is a change δQ in the charge beneath the gate where $\delta Q = \delta V C_g \simeq C_g \, \delta V_{gs}/(1 + j\omega C_g r_c)$; the relationship between δV and δV_{gs} follows directly from potential division by impedances, if the gate–drain impedance is high. The reader is now reminded of the relationship that a current I travelling for a time τ stores a charge $Q = I\tau$ in its path. Thus the change in charge δQ under the gate and the change in source–drain current can be related by $\delta I_{ds} = \delta Q/\tau$, where τ is the effective transit time through the channel under the gate. Consequently the relationship between I_{ds} and V_{gs} is given by

$$\delta I_{ds} = \delta V_{gs} g_{mo}/(1 + j\omega r_c C_g) = \delta V_{gs} g_m(\omega) \qquad 7.4$$

where $g_{mo} = C_g/\tau$. Thus in this simple treatment one finds an explicit relationship between the gate capacitance, transit time and mutual conductance at low frequencies. The value of g_{mo} can of course be obtained from the previous calculations at low frequencies. In general, with high-frequency FETs the fields along the channel are made sufficiently high so as to force the electrons to travel close to their scattering limited velocities v_s (see Chapter 8) around 10^5 m/s. Thus the transit time τ is the order of L/v_s where L is the channel length. The equivalent circuit can now be redrawn in the form of Fig. 7.5b.

At really high frequencies with relatively low gain so that voltage feedback is not important, the equivalent circuit can be simplified still further into that shown in Fig. 7.5c. This circuit allows one to calculate the maximum frequency for oscillation in the same manner as for the transistor. One notices that the power required at the input is $\frac{1}{2}i^2 r_c$ where $i = V_{gs} j\omega C_g/(1 + j\omega C_g r_c)$. With this interchange of i and V_{gs} it can be seen that the maximum power available from the output is then $P_{max} = i^2 g_{mo}{}^2 R_d/8\omega^2 C_g{}^2$ (using the maximum power transfer theorem). Thus the power gain available for the transistor is unity at f_{max}, given by

$$f_{max} = f_t/2(r_c/R_d)^{1/2} \quad \text{with} \quad f_t = g_{mo}/2\pi C_g = 1/2\pi\tau \qquad 7.5$$

It follows that one of the controlling influences in the high-frequency performance is the transit time of the charge carriers under the gate, and high-frequency structures must have narrow gate widths. In real devices there will also be significant resistance at the source and drain contacts. The source contact resistance R_s will essentially add to r_c and lower the frequency of maximum operation. The drain impedance is not so important unless it becomes comparable to R_d, the dynamic output impedance. As an example of the performance of a practical Schottky-gate FET, micrometre gate widths have given operation up to 18 GHz. The simplicity of construction and excellent definition of the evaporated metal gate make it likely that even better performances can be expected in the future.

7.3 The ideal metal–insulator–semiconductor (MIS)

One of the reasons for first looking at metal–insulator–semiconductor combinations came from the idea that one could change the character of a semiconductor from n-type into p-type by the application of an external field. Such a change is possible and leads to interesting devices. To demonstrate the ideas we shall start with a highly idealized model for the metal–insulator–semiconductor (MIS) combination shown in Fig. 7.6. Here it is assumed that the metal has a work function ϕ and the semiconductor an electron affinity χ such that the energy bands are flat in the absence of any applied potentials. The semiconductor shown here is n-type and, although

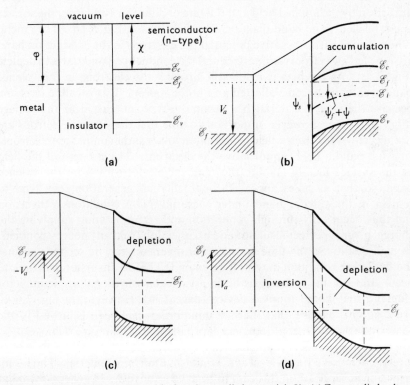

Fig. 7.6 An ideal MIS structure with changing applied potentials V_a. (a) Zero applied voltage. (b) Metal positive with accumulation of electrons. (c) Metal negative: depletion of electrons. (d) Metal negative: accumulation of holes.

all the arguments will be in terms of an n-type material, the reader should be able to transpose to a p-type material if required. The insulator represents a complete barrier to the motion of charge between the metal and semiconductor. In the real semiconductor–insulator interface there are usually surface charges but these are best left until later.

The insulator in this model prevents any electric current flow, so that both the holes in the valence band as well as the electrons in the conduction band are supposed to reach thermal equilibrium with the bulk of the semiconductor and the metal or insulator will play no part in supplying the charges that may be required when voltages are applied. Thus, even when a voltage is applied, the Fermi level in the bulk of the semiconductor is taken as the *reference level* for the energies of the holes and electrons at the edge of the semiconductor adjacent to the insulator. If, then, a positive potential is applied to the metal with respect to the semiconductor, electrons in the conduction band will be drawn to the surface of the semiconductor. An accumulation of charge will occur adjacent to the insulator until a new balance between the rate of attraction by the applied field and the rate of diffusion away from the surface is established. The band diagram then bends as in Fig. 7.6b. If the metal is now biased with a negative potential the electrons in the conduction band are repelled away from the interface and the conduction band starts to deplete (Fig. 7.6c). As the bias is made still stronger the electrons in the valence band eventually become affected and holes appear. This is called *inversion* because the n-type material has become p-type, at least near the surface (Fig. 7.6d). It is, however, important to note that the valence electrons are not affected by the electric field in the same manner as the conduction electrons. In a full band a field cannot move the electrons. The strength of the field at the surface merely favours the removal of electrons from the valence band, so that as holes appear through thermal generation they tend to remain in the valence band under these inversion conditions. Inversion can thus occur only through generation and recombination supplying the charge, given a perfect insulator and no injection from any nearby contacts. In high quality Si, the time constant for inversion may be seconds, so that the depletion condition may well be extended before inversion has time to occur. Indeed this is the basis of one particular functional semiconductor element, the charge-coupled device (page 174). However, in Fig. 7.6 all the diagrams assume that thermal equilibrium has been achieved in the semiconductor with any necessary lapse of time required to achieve it.

To discuss the effect mathematically it is convenient to define an energy level $\mathscr{E}_i$ such that a Fermi level at $\mathscr{E}_i$ would give intrinsic material. The Fermi level is then supposed in this case to be at $e\psi_f$ above $\mathscr{E}_i$. In general the band diagrams bend and the local electric potential in the conduction band is given by ψ, with $\psi = 0$ in the bulk of the material. The Boltzmann approximation of eqns 3.21, 3.22, and 3.24 can be rewritten as

$$n = n_i \exp\left[e(\psi_f + \psi)/kT\right] \qquad p = n_i \exp\left[-e(\psi_f + \psi)/kT\right] \qquad 7.6$$

Electrical neutrality in the bulk material, assuming that there are N_d ionized donors, implies that

$$N_d = n_i \exp\left(e\psi_f/kT\right) - n_i \exp\left(-e\psi_f/kT\right) \qquad 7.7$$

Gauss' theorem gives

$$\varepsilon_0 \varepsilon_r \frac{dE}{dx} = e(N_d - n + p) \qquad 7.8$$

with $E = -d\psi/dx$.

The integration can be partially performed if one notes that $d^2\psi/dx^2 = (\frac{1}{2}d/d\psi)(d\psi/dx)^2 = (\frac{1}{2}d/d\psi)(E^2)$; then, from eqns 7.6 to 7.8 and putting in the boundary condition that $E = 0$ at $\psi = 0$, the integration yields

$$\tfrac{1}{2}\varepsilon_0 \varepsilon_r (E_s)^2 = -kT N_d \left[(e\psi_s/kT) - \frac{\cosh e(\psi_f + \psi_s)/kT}{\sinh e\psi_f/kT} + \coth e\psi_f/kT \right] \qquad 7.9$$

where E_s is the field at the surface and ψ_s is the surface potential. Note that a positive surface potential implies that the bands bend downwards. Figure 7.7 plots an example of the variation of the surface field E_s against the surface potential appropriate to a n-type extrinsic semiconductor. There are three distinct cases to be considered:

(a) *Accumulation.* When the surface potential is positive the electrons are attracted to the surface and the middle term in the bracket of eqn 7.9 dominates. A very small change in surface potential requires a very large change in the electric field at the surface.

(b) *Depletion.* As the potential ψ_s decreases into the range $0 > \psi_s > -2\psi_f$ so the n-type carriers fall well below their equilibrium value at the surface, and in effect the region close to the surface behaves as a conventional depletion region of width W, with

$$|\psi_s| \sim \tfrac{1}{2}eN_d W^2/\varepsilon_0\varepsilon_r \quad \text{and} \quad E_s \sim eN_d W/\varepsilon_0\varepsilon_r$$

The first term in 7.9 is then the important one.

(c) *Inversion.* As the surface potential decreases still further so eventually $|\psi_f + \psi_s| > \psi_f$, and the middle term of eqn 7.9 again dominates though ψ_s is negative. The material has inverted at its surface where an accumulation of holes has formed. It must be emphasized that this accumulation of holes occupies a very thin layer (typically a Debye length or two), much thinner than a depletion width of the n-material. The accumulation of holes shields the depletion region from further widening and so the depletion is effectively

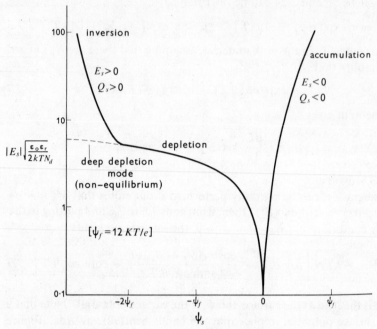

Fig. 7.7   Surface charge or field, in ideal MIS, against surface potential ψ_s.

fixed at its maximum width, which will occur around $\psi_s \rightarrow -2\psi_f$, with $W_{\text{max}} \sim 2(\psi_f \varepsilon_o \varepsilon_r / eN_d)^{1/2}$. If, however, the surface potential is changed rapidly compared to the time required to generate the holes at the surface, then the depletion region will continue to widen and the surface potential will change by a larger amount for a given change in field at the surface. This is shown as the dashed line in Fig. 7.7.

Before leaving this calculation the reader's attention is drawn to the fact that use of the Fermi level and the voltage ψ have automatically balanced the field induced and the diffusion currents in a single equilibrium calculation. Nevertheless, when there is an accumulation of charge at the surface the correct physics is that the charge is drawn to the surface by a field and diffuses away from the surface, the two currents of diffusion and field-assisted motion cancelling.

Differential capacitance

Considerable information about real metal–insulator–semiconductor (MIS) combinations can be obtained by making measurements of the capacitance between the metal and semiconductor. This measurement of the overall capacitance C/unit area is made by small changes of voltage about a steady bias voltage across the MIS structure: the capacitance so measured is called the differential capacitance. To avoid the continual insertion of the term *per unit area* it will be understood that in the following discussion a unit area is being considered.

The charge stored in the semiconductor Q_s will cause it to act like a capacitance given by the positive quantity $C_s = -dQ_s/d\psi_s$, at least for small changes in ψ_s. This capacitance C_s will then act in series with the insulator capacitance C_i $(= \varepsilon_o\varepsilon_i/W)$ to give the value of $C = C_sC_i/(C_s + C_i)$ as the MIS capacitance. A proof of this follows.

If E_i is the field in the insulator then the voltage across the insulator is $-E_iW_i$ and the total voltage across the MIS structure is $V = \psi_s - E_iW_i$. The value of E_i can be found from Gauss' theorem as $\varepsilon_0\varepsilon_iE_i = Q_s$, and similarly the charge on the metal is $Q = -Q_s$. The differential capacitance C is given by $(1/C) = dV/dQ = -dV/dQ_s$ and substituting for V gives $(1/C) = -d\psi_s/dQ_s + (W_i/\varepsilon_o\varepsilon_i)$. Hence $(1/C) = (1/C_s) + (1/C_i)$, showing that the insulator and semiconductor capacitances act in series as stated.

The semiconductor capacitance C_s is plotted in Fig. 7.8 for the example considered, with $\psi_f = 12kT/e$. The reference capacitance C_{so} is the capacitance $(e^2N_d\varepsilon_0\varepsilon_r/kT)^{1/2}$ per unit area that is appropriate at $\psi_s = 0$. (This can be shown analytically by a series expansion of eqn 7.9 for small values of ψ_s.) The minimum capacitance occurs approximately at the point where inversion

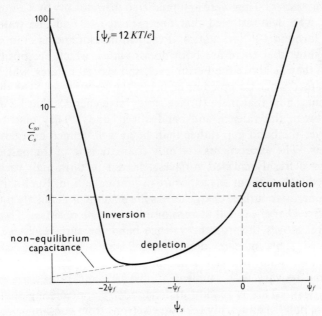

Fig. 7.8   Semiconductor capitance against surface potential.

starts (i.e. $\psi_s \simeq -2\psi_f$); the maximum depletion depth is then W_{max}, so that $C_{min} = \varepsilon_0\varepsilon_r/W_{max}$ (per unit area). The minimum value of C, the actually measured value, is of course modified by the insulator capacitance C_i, as would be expected.

The dashed curve in Fig. 7.8 has again been inserted to emphasize the point that minority carriers take time to be generated. If the measurement of the differential capacitance across the MIS structure is made at a frequency of shorter period than the time required for readjustment of the minority carriers, then the accumulation of minority carriers will not play a role in determining the capacitance. It will only be the majority carriers that will change, and the appropriate terms must then be omitted for the calculation of $dQ_s/d\psi_s$, the omitted terms being the minority charges in Q_s. In such a measurement the steady bias field E_s will settle to the value given by the solid line in Fig. 7.7.

7.4 The real surface

At the surface of a freshly cleaved piece of semiconductor there are usually dangling valence bonds from the valence electron sites which are not paired with electrons as they are in the bulk. The discontinuity in the electron binding gives extra states for the electrons over and above those in the bulk: these are *surface states*. They were first predicted theoretically by Tamm and by Shockley, who demonstrated that one expected to find one state per surface atom (around $10^{19}/m^2$ states). Practical measurements[7.5] in *ultra-high vacuum* show that there are both donor states, which are positively charged when they lie above the Fermi level, and acceptor states, which are negatively charged when occupied or lie below the Fermi level. The theory and experiments show that these surface states have energies that lie in the band gap between the valence and conduction bands. They tend to be distributed over the band gap rather than lie at well defined energies like impurity levels. The experiments are only characteristic of the particular semiconductor if freshly cleaved surfaces, cleaved in ultra-high vacuum, are used immediately without any exposure to contamination. Such surfaces tend to have a negative surface charge unless the material is definitely p-type, when the surface charge can fall to zero or even become positive. Thus, as the Fermi level moves towards the valence band the surface charge can become positive. This indicates donor surface states close to the valence band but acceptor states nearer the middle of the band gap. The surface charge is found to vary with ambient atmosphere and processing in an unpredictable way. This is readily understood by appreciating that the broken valence bonds can readily take up electrons from contamination or gas molecules falling on the surface. Thus the real surface has always been difficult to study and to characterize. Surface states can enhance recombination; the net surface charge can also cause inversion. For example, the negative surface charge on n-type material can be so large that inversion has to occur so as to ensure overall electrical neutrality. The surface then becomes p-type. This can enhance recombination and increase leakage currents across p-n junctions, thereby limiting the performance of many practical devices.

For silicon, a dramatic change is brought about by growing silicon dioxide over the surface as a protective layer. In one of several methods this is done in a hot, moist oxygen atmosphere. It is found that the 'dangling bonds' are incorporated into the SiO_2 and the density of surface states is reduced by orders of magnitude. Indeed the $Si-SiO_2$ interface has a positive rather than negative surface charge, indicating that the surface states are donors. The total density of surface states is then reduced to the order of $10^{15}/m^2$ and these are found to be spread over the band gap. With careful preparation, the number of states is independent of ambient atmosphere and can be stable with time. Inversion and surface recombination are reduced, and moreover the SiO_2 can act as a mask against the diffusion of impurities. When a diffusion is made through a window of SiO_2 it will proceed under the oxide as well as down into the material. By this means the actual metallurgical junction is protected from the atmosphere. All these benefits go into the success of the 'Planar' process of construction which uses the $Si-SiO_2$ interface properties as outlined in Chapter 2.

The thermal growth of silicon dioxide is not the only method of protecting the surface. With the technique of sputtering,† almost any insulator can be deposited on a semiconductor and its merits assessed. Silicon nitride (Si_3N_4) is another currently useful insulator which has greater resistance to diffusion of certain impurities (e.g. sodium). However, whatever the insulator, the basic aim must be to stabilize the unbound valence electrons at the interface of the two materials.

To describe the surface more mathematically, it is supposed that the surface states are arrayed continuously over the band gap with some density N_{ss}/m^2 per unit energy range. Indeed practical measurements[7.6] show that in the middle of the band gap the density of surface states, N_{ss}, is more or less constant. In the $Si-SiO_2$ interface these states are donors, so that they are positively charged if they are unoccupied (that is they lie above the Fermi level at the surface). Now with a positive surface potential, the Fermi

† When an electrical discharge is formed in a gas under a low pressure (e.g. neon signs) it is found that the heavy, positively charged, gas ions have considerable energy. On hitting material placed in the discharge these ions can knock out atoms of the material. These *sputtered* atoms can then travel in the low-pressure gas and deposit themselves on adjacent material. An inert gas is commonly used for the discharge. Sputtering can also be used to clean a surface prior to deposition of other material.

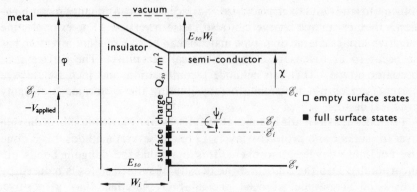

Fig. 7.9   MIS structure, showing surface states and applied voltage for flat band condition.

level rises (Fig. 7.6b) giving less of the positively charged surface states. Thus if the surface potential is ψ_s, then the surface state charge is (per unit area)

$$Q_{ss} = Q_{so} - e \int_0^{\psi_s} N_{ss} \, \mathrm{d}\psi_s \qquad 7.10$$

where Q_{so} is the charge that exists when the bands are flat, i.e. when the surface potential is zero. It should be noted how the Fermi-Dirac distribution has been approximated by its zero temperature form of Fig. 3.4b. To achieve the flat band condition there must be a field in the insulator given by $E_{io} = Q_{so}/\varepsilon_o\varepsilon_i$ (from Gauss' theorem). This will then require the metal to be at a potential $\{-W_iE_{io} - \chi + \phi - (\mathscr{E}_c - \mathscr{E}_f)/e\}$ with respect to the semiconductor (see Fig. 7.9). In this expression, allowance has now been made for a more realistic metal than in Fig. 7.6a but no allowance has been made for possible charges in the insulator. If the surface states are very high in density then the potential $E_{io}W_i$ will dominate, and it may be very difficult or even impossible to achieve a flat band condition without too high fields appearing and causing breakdown. As the surface potential is changed, so the surface state charge will change, and larger variations in the applied potential will be required (to force changes inside the semiconductor) than those required for the ideal MIS structure. The low density of surface states is thus very important practically if useful changes of charge are to be made inside the semiconductor.

There are several clever methods of measuring the density of surface states; one of the easiest to understand uses the ideal MIS model as a guide. Charges in the insulator are assumed negligible, but the surface state charges Q_{ss} and the semiconductor charge Q_s are lumped together to define a semiconductor capacitance $C_s^* = -\mathrm{d}(Q_{ss} + Q_s)/\mathrm{d}\psi_s$ per unit area, instead of $C_s = -\mathrm{d}Q_s/\mathrm{d}\psi_s$ as for the ideal MIS. The voltage across C_s^* is ψ_s and the voltage across the total capacitance C is V, the applied potential. As for the ideal MIS, the capacitances C_s^* and C_i are in series to form the measured differential capacitance (per unit area always understood) C. Thus $C_s^* = CC_i/(C - C_i)$; so that C_s^* can be recovered from knowledge of C. However, from the potential division, $\delta V(C/C_s^*) = \delta\psi_s$, so that one can integrate the expression for $\delta\psi_s$ from the differential capacitance measurements at different bias voltages V:

$$\psi_s(1) - \psi_s(2) = \int_{V_1}^{V_2} \left[1 - (C/C_i)\right] \mathrm{d}V \qquad 7.11$$

And from the definition of C_s^* one can see that

$$C_s^* = eN_{ss} + C_s = CC_i/(C - C_i) \qquad 7.12$$

Hence, by using the results for C_s from the ideal MIS, eqns 7.11 and 7.12 enable one to find the surface states as a function of the surface potential. Usually additional methods are required to augment this evaluation [see Problem 7.3]. Figure 7.10 sketches a typical variation of the surface states in the Si–SiO$_2$ interface system. In the middle of the band gap N_{ss} is found to be reasonably constant, but peaks are found closer to the band edges.

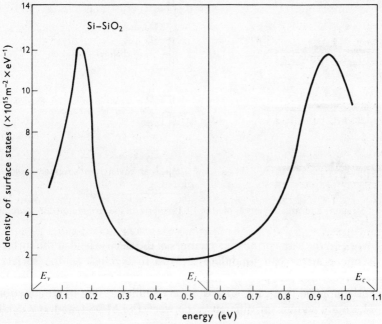

Fig. 7.10 Typical surface-state density of an Si—SiO$_2$ (after S.M. Sze, reference 1.5, with permission).

In conclusion, it must be noted that the surface states give rise to surface charges which change with surface potential. Only by good control of these surface states and keeping their density low can one control the charge inside the semiconductor in the manner approaching the ideal MIS. Charges stored *in* the insulator must also be kept low, or at least tightly controlled, if reliable and repeatable results are to be obtained. Control of the surface conditions has at least proved possible in the Si–SiO$_2$ interface, and we shall now discuss the applications.

7.5 The metal–oxide–semiconductor transistor (MOST)[7.8,7.9]

The ability to grow consistent charge-free layers of SiO$_2$ over Si has lead to the growth and practical use of the MOST (metal–oxide–semiconductor transistor†). Four basic forms for this device are sketched in Fig. 7.11 and

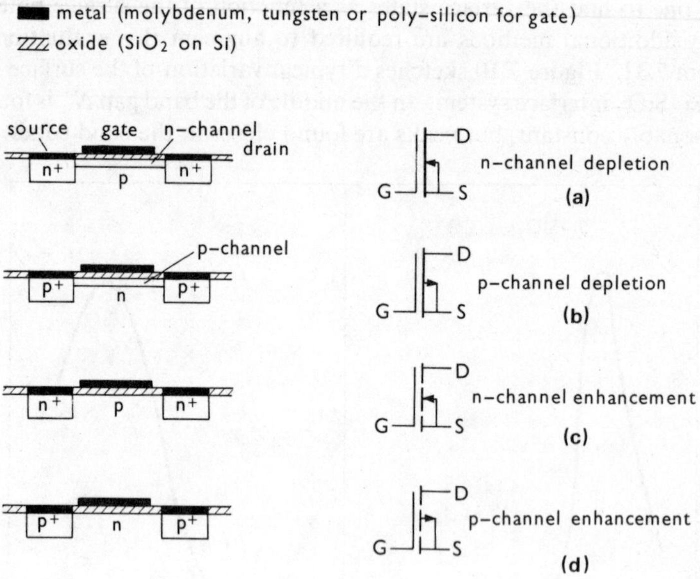

Fig. 7.11 Structure and circuit symbols of MOSFETs (substrate shown connected to source in circuit diagram).

they all work on the same principle: changes of the electric field at the surface of the semiconductor can modulate the resistivity close to the surface. Thus the current that can pass between a source and a drain is controlled by the potential applied to a gate, an action very similar to that of the junction FET. There are significant differences between the MOST and JUGFET

† Also called MOSFET (metal–oxide–semiconductor field effect transistor), MISFET (metal–insulator–semiconductor FET) and IGFET (insulated-gate FET).

because in the JUGFET the gate can only be satisfactorily biased in one direction. In the MOST, the gate can either attract or repel the charge carriers in the channel beneath without any gate current flowing. Thus one can start with an n-channel (Fig. 7.11a) and repel the electrons beneath the channel and form a *depletion* MOST (working analogously to the JUGFET) or one can have the n-channel formed by inversion of a p-type substrate (Fig. 7.11c). This makes an *enhancement* MOST. In a good enhancement device there is no conduction when the gate−source voltage is zero, conduction only occurring when the gate−source voltage exceeds some *threshold* voltage V_{th}. There are, of course, the complementary p-channel devices. The enhancement MOST is particularly useful in low-power, high-packing-density logic circuits for use in large-scale integrated circuits with computer applications. The discussion will concentrate on this type of MOST.

Because the surface potential at the Si−SiO$_2$ interface tends to be positive, n-type material tends to have a lower resistivity at its surface than in its bulk while p-material tends to invert even with no field in the oxide. Thus enhancement MOSTs, with zero conduction at zero gate bias, are more easily made in p-channel form by an inversion of n-material at a negative threshold voltage on the gate. It will be this type that we discuss. Figure 7.12 indicates a schematic diagram for an idealized model that will give the general form of the characteristics over a limited range. In this model it is assumed, first, that inversion will start when a potential $-V_{th}$ appears across the gate-channel region, and second, that at the inversion point the surface charge is Q_{so} per unit area, this value combining surface state and semiconductor charge. This charge is not mobile because it is either ionized impurities in the depletion region or surface-state charge. To create mobile charge, the gate voltage has to be increased in magnitude to a value $-V_{gs}(V_{gs} > V_{th})$. Inversion is then created and the inversion charges are mobile. From Fig. 7.6 it can be seen that a large change in surface field (and hence inversion charge) can occur with a small change in surface potential. The change in surface state charge can then be small provided also that the density of surface states is low. Hence, in this simplified model, the

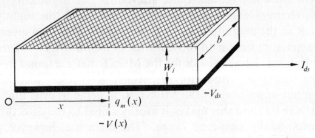

Fig. 7.12 Notation for MOS theory.

surface-state charge is supposed to be a constant. The extra field provided by a *gate–channel* voltage $-V_{gc}$, over and above the value $-V_{th}$, gives a mobile charge, Q_m/unit area, of holes such that $Q_m = (V_{gc} - V_{th})\varepsilon_o\varepsilon_i/W_i$, directly from Gauss' theorem. If the potential of the channel under the gate is now $-V$ with respect to the source, then $V_{gc} = V_{gs} - V$, and so the actual mobile charge/unit distance in the breadth b is

$$q_m = (V_{gs} - V - V_{th})\varepsilon_o\varepsilon_i b/W_i \qquad 7.13$$

Thus the current flow at any point is given by the field $-\mathrm{d}V/\mathrm{d}x$ driving the charges with an effective mobility μ_p. The effective mobility at a surface will be influenced by a number of factors. At a good surface where the density of surface states is low, it will primarily be the scattering of the electron wave at the surface of the material which will reduce the mobility from the bulk value. At the Si–SiO$_2$ interface the mobility can be reduced by as little as a factor of three from its bulk value. If the density of surface states is very high, one can find that the expected mobile electrons become trapped in these surface states and the mobility is reduced still further. Allowing for an effective mobility μ_p, the current I at all points is given by:

$$I = (\mu_p\varepsilon_o\varepsilon_i b/W_i)(\mathrm{d}V/\mathrm{d}x)(V_{gs} - V - V_{th}) \qquad 7.14$$

The current is everywhere equal to the current (I_{ds}) flowing between drain and source so that, provided inversion occurs $(V_{ds} < V_{gs} - V_{th})$ along the whole channel, one may integrate to find the characteristic for the *triode* region in the same way as in the calculation for the JUGFET

$$I_{ds} = (\mu_p\varepsilon_o\varepsilon_i b/WL)[(V_{gs} - V_{th})V_{ds} - \tfrac{1}{2}V_{ds}^2] \qquad 7.15$$

This gives the relationship between the current I_{ds} and source–drain voltage V_{ds} in the 'triode' region. Whenever $V_{ds} > V_{gs} - V_{th}$ then the mobile charge in the accumulation is reduced to zero and inversion no longer occurs over the entire channel. This is analogous to 'pinch off' in the depletion FET. One notes that at the limit $\mathrm{d}I_{ds}/\mathrm{d}V_{ds} = 0$ for constant V_{gs}, and the current saturates. In practical diodes, for similar reasons as in the JUGFET, the current does not remain quite constant. The arguments presented there about channel-length modulation still hold, and this slightly increases the current as the magnitude of V_{ds} is increased. This then gives the saturation characteristics similar to those for the depletion FET. Figure 7.13 sketches the idealized characteristics for the MOST that are found from this theory.

The equivalent circuit for the MOST is virtually identical to the FET. The input impedance of the MOST is, in general, even higher than that of the JUGFET, and this makes it ideally suited to the detection of very small currents, in the pico-amp range. The main use, however, is probably in large-scale integrated memory circuits. In this application it is important to have the threshold voltage V_{th} accurately controlled at values around 1 to 1·5 V. The potential of the substrate with respect to the source can be changed from its zero value given in the theory above, and this can give additional control over the value of V_{th}. Nevertheless, to bring the values into the correct range, great care has to be taken to exclude unwanted charges from the oxide layer. Sodium ions in the oxide have been found to cause trouble in giving unstable values of V_{th}, and great care has to be taken

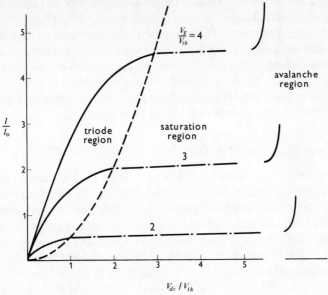

Fig. 7.13   MOSFET enhancement characteristics. $V_{th} \sim 1·5$; $I_0 = \mu_p \varepsilon_0 \varepsilon_i V_{th}^2 (b/W_i L)$.

to exclude sodium impurities. The impurity densities of the semiconductor material must be closely controlled, and the gate metal must be suitably chosen so that its work function does not cause a large shift in the effective value of V_{th}; tungsten, molybdenum and poly-crystalline silicon all find use as suitable materials for gates. In handling MOS devices one must remember that electrostatic charges can readily destroy the device: since $q = VC$, charges on small capacitances can give rise to voltages large enough for breakdown. Protective Zener diodes (see Chapter 8) are sometimes incorporated into the device across the gate to substrate. With these diodes only a few volts can appear across the gate.

A new addition to the MIS family was found when silicon nitride (Si_3N_4) was used as an additional insulating layer over the SiO_2 in an MIS structure. It is believed that Si_3N_4 is more resistant to diffusion of ions, especially sodium ions. Uncertain amounts of charged ions in the insulating layer create uncertain electric fields between the gate and the semiconductor, and this in turn creates a variable turn-on or threshold voltage for the MIS transistor. Thus a layer of Si_3N_4 should prevent ion diffusion and lead to greater control over the gate voltage. However, it was found that such a layer created *interface states* that could trap *electrons* at the Si_3N_4–SiO_2 boundary. These interface charges could be controlled by the application of short high-voltage pulses (around ± 25 volts for 100 μs) which presumably allowed electrons to tunnel through the insulating layers to populate the interface states. These charges, like any in the insulator, affect the field at the surface of the semiconductor. Thus the gate voltage required to turn on the channel in the semiconductor will be controlled by the interface charges. Provided these interface charges are stable and non-volatile, one has a new element with a controllable threshold voltage (say from -1 to -8 V). These MNOS (metal–nitride–oxide–silicon) transistors should add greatly to the versatility of MIS devices in their use for computers.[7.14]

7.6   Some uses of the MOST in ICs

One of the major uses of the MOST lies in the IC field of random-access computer memories where small size and low power consumption are important. One of the well-tried circuits is the flip-flop memory element shown in Fig. 7.14*a*. Its action should be considered in terms of two coupled NOR (neither-nor) circuits. Consider first the single NOR circuit formed by T_1, T_2 and T_3. The transistor T_3 is acting as a load resistor (often over 100 k ohms) to the transistors T_1 and T_2. If either A or B inputs activate these transistors then the voltage at E will fall close to zero (the logical 0 value). Alternatively if neither A nor B is activated then both T_1 and T_2 remain firmly off and the voltage at E is close to V_{dd} (the logical 1 state). Thus if neither A nor B is at logical 1 then E is at logical 1. Coupling the two NOR gates together forces either E or F to be at logical 1 and the other at logical 0. The inputs A and D activate the required state and so six devices form a memory element that remains in a given state until changed.

The switching time is primarily determined by the rate at which current from the MOST can charge the gate input capacitances C_g. From the equivalent circuit for the FET one can see that the current generator has its strength determined by the value of g_m, the mutual conductance. Thus the time constant is the order of $\tau = C/g_m$. In spite of very low currents, the capacitances are so small that switching times in the 100 ns range are currently possible. The power dissipation can be reduced by having V_{gg} attached to the clock pulse of the computer, so that the cell draws power only periodically. The information stored in the cell is not in fact lost because the gate capacitances have such a low leakage that they retain their voltage for at least a clock cycle. As soon as the devices are reactivated the gates are replenished in charge. Power dissipations can be kept well below the milliwatt per bit when the devices are operated in this 'standby' mode.

A higher packing density can be achieved by making full use of the charge-storage properties of the gate capacitance. One possible scheme is shown in Fig. 7.14*b*, where only three MOSTs per bit are required. The action of this circuit is that whenever the *write enable* line switches-on T_1 the capacitance of the gate at T_2 (dotted for emphasis in the figure) becomes charged to the

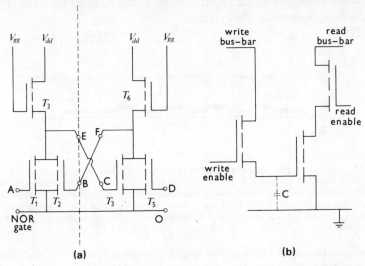

Fig. 7.14 MOS memory cells. (*a*) Combination of two NOR gates (flip-flop). (*b*) Refresh type of memory cell.

value of the *write bus bar* voltage. The write action is done periodically in time with the clock pulse of the computer so as to refresh the charge on the gate of T_2, the memory cell. The *read enable* line allows a current through the read bus bar and T_3, which can sense the state of T_2 and hence the state of the memory. Other, but comparable, arrangements can be made. For current technology a useful order of magnitude is that a 'bit' occupies around 100 μm^2 and consumes about 100 μW.

Even higher packing densities are promised for the future with the *charge-coupled* memory element (Fig. 7.15). If the oxide is grown on Si at low temperatures, very low recombination rates have been found possible because there are fewer dislocations. The time scale of generation and recombination at the surface can then be several seconds, so that inversion does not occur on the time scale of a computer's clock-pulse cycle. A gate that is biassed into a region where inversion would be expected then has only a depletion region, though of deeper extent than usual. This is called a *deep depletion mode* of operation. The presence or absence of holes under the gate can then be used as a basic memory element. Figure 7.15 shows schematically a shift register using these ideas. The presence or absence of holes in the first cell C_1 is determined by the gate input, V_{gate}, the holes being derived from a source S. The other cells are driven periodically into and out of a deep depletion state. The potential wells formed by the deep depletion are designed to favour the movement of holes, towards the drain, from cell to cell, as the potentials V_1 and V_2 are cycled about V_0 (see Fig. 7.15). Finally the charge is read out at the drain.

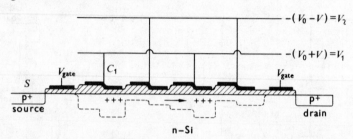

Fig. 7.15 Charge-coupled shift register in schematic form. Shaped electrodes force field to favour motion of holes towards drain. Changing V to $-V$ cyclically forces this motion from one cell to the next. The spacing between the electrodes has to be small ($\sim 2\ \mu$m) so that the charge is not lost on transferring from one cell to another by falling into 'traps' caused by the surface states and is not lost by diffusion. Efficiency of transfer can be almost 100% with such small spacings.

The generation of holes at the inversion region can be accomplished by methods other than injection from a source. Optical generation of holes (see Chapter 9) can turn the device into a picture-tube, with the charge being proportional to the incident light. Avalanche breakdown can also generate the holes required, so that simply pulsing the gate voltage into avalanche state could feed in the holes when required. The device and the techniques are at present in the research stage but show considerable promise.[7.11,7.12]

PROBLEMS

7.1 Show from eqn 3.13a that

$$J_n \int_{-d}^{0} \exp -(eV/kT)\ \mathrm{d}x = eD_n\ |n\ \exp\ (-eVkT)|_{-d}^{0}$$

where V is the potential in the conduction band and d is the depletion width of the Schottky barrier.

Take $V = 0$ at $x = 0$, the metal–semiconductor interface. Take ϕ_b to be the barrier and $V = \phi_b - \delta E_f - V_a$ the potential at $x = -d$. The carrier concentration in the bulk of the semiconductor is $n = N_d = N_c \exp -\delta E_f/kT$. The depletion width is given from $\frac{1}{2}eN_dd^2 = \varepsilon_o\varepsilon_r(V_0 - V_a)$ where $V_0 = \phi_b - \delta E_f$. Approximate to the integral with a potential $V = -(eN_ddx/\varepsilon_o\varepsilon_r)$ forgetting the quadratic term because of the rapid decay in value of the integrand.

Hence show that

$$J_n \simeq e\mu_n N_c (eN_d \, d/\varepsilon_0 \varepsilon_r) \exp\left(-e\phi_b/kT\right)\left[\exp\left(eV_a/kT\right) - 1\right]$$

gives the approximate diffusion current density through a Schottky barrier.

If $V_0 - V_a \sim \phi_b \sim 1\,\text{eV}$, $N_d \sim 10^{22}/\text{m}^3$, $\mu_n \sim 0.1\,\text{V}\cdot\text{s/m}^2$, $N_c \sim 10^{25}/\text{m}^3$, $\varepsilon_r \sim 12$, then compare this diffusion current density with the thermionic current density at room temperature (and with $m^*/m \sim \frac{1}{2}$). Thus show that the device will have its current limited by the thermionic emission density rather than the diffusion current density.

7.2 Show that the attractive force between a metal and an electron is $e^2/16\pi x^2 \varepsilon_0 \varepsilon_r$. The field close to the metal may be taken as approximately constant at its peak value E_p so that the reduction in potential at a distance x from the metal is then $[E_p x + e/16\pi x \varepsilon_0 \varepsilon_r]$. Find the point where this potential reduction is a maximum and hence show that the reduction in the barrier height for a Schottky diode is $(eE_p/16\pi\varepsilon_0\varepsilon_r)^{1/2}$ volts. Estimate this potential reduction caused by the image force (referred to as *Schottky effect*) for a material where the donor density is of order $10^{22}/\text{m}^3$ with $\varepsilon_r = 12$ and the voltage across the depletion region is of the order of half a volt. [Answer: approx. 20 mV]. [Hint: use method of images, ref 3.9.]

7.3 Capacitance measurements of surface states are not always easy to interpret because of the masking effect of the oxide capacitance and the accuracy required. Nicollian and Goetzberger[7.6] measure the losses required to store and unstore charge on the surface states. Figure 7.16 shows an equivalent circuit envisaged at each bias level. The components are differential components, i.e. for small changes of voltage about a bias value. The oxide capacitance C_{ox} does not vary with bias and is measured under strong accumulation conditions.

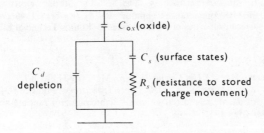

Fig. 7.16

Find an expression for the impedance of the equivalent circuit. Subtract the oxide impedance $-j/\omega C_{ox}$. Express the remaining impedance as an admittance and show that the real part of this admittance is $G = C_s \omega^2 \tau/[1 + (\omega\tau)^2]$ where $\tau = C_s R_s$ is the time constant for charging and discharging the surface states. Show that G/ω has a maximum when $\omega\tau = 1$ and this maximum is C_s. Hence outline a possible method for calculation of the surface states.

7.4 Sketch the band diagram for a Schottky barrier where there is a very thin 'oxide' layer between the metal and the semiconductor. Assume that the density of surface states at the surface of the semiconductor is so high that a very small movement of the Fermi level away from some energy E_{fo} at the surface creates so much surface charge that any field across the 'oxide' can easily be accommodated with a negligible movement of the Fermi level. Show that in this case the Fermi level is pinned close to the energy E_{fo} and that the barrier height for the diode is close to $(E_c - E_{fo})$ regardless of the work function of the metal or the electron affinity of the semiconductor. It is assumed that there is no surface charge if E_f passes exactly through E_{fo}.

7.5 Explain why a Si Schottky barrier across the collector–base of a Ge transistor does not protect the transistor against charge storage in the saturation region.

7.6 Take the values for a MOSFET as $b = 100\,\mu\text{m}$, $W = 1\,\mu\text{m}$, $L = 10\,\mu\text{m}$, $V_{th} = 1.5\,\text{V}$, $\varepsilon_i = 3.9$, $\mu_p = 0.06$ and calculate I_0 in Fig. 7.13.

General references

GROVE, A. S. Reference 2.10.
7.1 SZE, S. M. Reference 1.5.
    Particularly good set of references on Schottky barriers.

Special references

Schottky Gate FET

7.2 DRANGEID, K. E. and SOMMERHALDER, R. Dynamic performance of Schottky barrier field effect transistor. *IBM J. Res. Devlpmt*, **14** (1970), 82–94.
    See also other articles in same month's edition.
7.3 HOOPER, W. W., CAIRNS, B. R., FAIRMAN, R. D. and TREMERE, T. A. The GaAs Field Effect Transistor. In *Semiconductors and Semimetals*, ed. R. K. Willardson and A. C. Beer. Vol. 7, pt A. Academic Press, 1971.

MIS etc.

7.4 LAMB, D. R. *Electrical Conduction Mechanisms in Thin Insulating Films*. Methuen, 1967.
7.5 ALLEN, F. G. and GOBELLI, G. W. Work function, photoelectric threshold and surface states of atomically clean silicon. *Phys Rev.* **127** (1962), 150–158.
7.6 NICOLLIAN, E. H. and GOETZBERGER, A. The Si–SiO$_2$ interface electrical properties as determined by the MIS conductance technique. *Bell System Tech. J.* **46** (1967), 1055–1133.
7.7 KOOI, E. The surface properties of oxidised silicon. Philips Technical Library, 1967.

MOSFETs

7.8 CRAWFORD, R. H. *MOSFET in Circuit Design*. McGraw-Hill, 1967.
7.9 WALLMARK, J. T. and JOHNSON, H. *Field Effect Transistors*. Prentice Hall, 1966.
7.10 VADASZ, L. L., CHUA, H. T. and GROVE, A. S. Semiconductor random-access memories. *IEEE Spectrum*, May 1971, 40–48.

Charge-coupled devices

7.11 BOYLE, W. S. and SMITH, G. E. Charge Coupled Devices—a new approach to MIS device structure. *IEEE Spectrum*, July 1971, 18–27.
7.12 TOMPSETT, M. F. Charge transfer devices. *J. Vacuum Sci. Technol.* **9** (1972), 1166–1181.

Barrier transit time diodes

7.13 COLEMAN, D. J. and SZE, S. M. A low noise metal–semiconductor–metal (MSM) microwave oscillator. *Bell System Tech J.* **50** (1971), 1695–99.

MNOS transistor

7.14 FROHMAN BENFCHKOWSKY, D. The metal–nitride–oxide–silicon (MNOS) transistor, characteristics and applications. *Proc. IEEE*, **58**, (1971), 1207–19.

8

Hot Electron Devices

8.1 Hot electrons

A hot crystal has a higher mean thermal energy of vibration among its atoms than a cold crystal. Similarly a hot gas has a higher level of random kinetic energy among its particles than a cold gas (in a classical gas obeying Boltzmann statistics the mean kinetic energy per particle is $\frac{1}{2}mv^2 = \frac{3}{2}kT$). When an electric current is passed through any conductor or semiconductor, the charge carriers gain kinetic energy from the applied electric field. Collisions with the crystal lattice redistribute this energy randomly among the charge carriers and also give up some of this energy to the lattice, increasing its vibrational energy. Thus both the gas of charge carriers and the crystal lattice become hotter. Now the conduction of heat in any solid is accomplished by two methods: the transport of thermal vibrational energy by the lattice waves (discussed in section 6.7) and the transport of thermal kinetic energy by the mobile electrons. In a metal, there are so many electrons that the latter process often dominates and K, the thermal conductivity of the metal crystal, is closely linked to its electrical conductivity, σ. For many metals at not too low temperatures, one can show that $(K/\sigma T) \sim 2\cdot45$ watt ohm/deg^2. This relationship is known as the Wiedemann-Franz law.[2.1] One then finds, in a metal, that the lattice temperature and the temperature of the electron gas are essentially the same. However, in a semiconductor there can be far fewer charge carriers, so that the lattice vibrations rather than the electrons determine the conduction of heat. One says that the heat is conducted by phonons. If such a semiconductor is attached to a heat sink, then it is possible to keep the lattice relatively cold while allowing the electrons to gain considerable energy from an electric field. For applied fields below 10^5 V/m the increase in kinetic energy caused by the field is relatively small compared to the thermal equilibrium energy at room temperature (indeed this was precisely the assumption in calculating the current flow in a p-n junction). However, if the fields are sufficiently high, the charge carriers have an energy that is much greater than if they were in thermal equilibrium with the crystal lattice. Such carriers are said to be *hot*. Hot carriers behave differently from those at lower energies. The

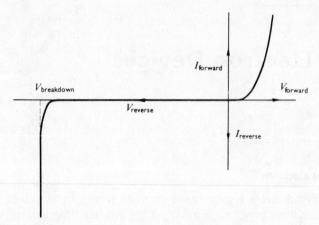

Fig. 8.1 Voltage-current characteristics as breakdown of a p-n junction.

concept of mobility breaks down, the velocity tending to be constant, or even decrease, with an increasing electric field. Hot carriers can liberate more charge carriers through energetic collisions with the valence electrons of the host atoms. This latter effect gives rise to *avalanche breakdown*. A p-n junction under reverse bias can exhibit avalanche breakdown, when the current magnitude can change over a decade or more with a minimal change of the voltage away from a 'breakdown' value (Fig. 8.1). Such a diode is called a *voltage stabilizer* diode or, often, a Zener diode. These diodes are used for giving a reference voltage.[8.5]

The search for a semiconductor device that can generate reasonable powers at microwave frequencies has been a long one. Devices using hot-carrier effects are meeting some of these needs. For example, spontaneous oscillations of current and voltage build up across certain types of p-n junction when these are biased to avalanche breakdown and placed in resonant circuits equivalent to the type in Fig. 8.2. The reader must appreciate that at very high frequencies (10^{10} Hz) the structure of resonant circuits is quite different to that at low frequencies, though the analysis is mostly done in terms of conventional elements indicated in Fig. 8.2. These *avalanche diode oscillators* are discussed in section 8.4.

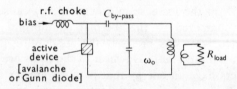

Fig. 8.2 Oscillator circuit (schematic).

The final device, discussed in this chapter, is structurally even simpler than the avalanche diode in that there is no p-n junction—it is merely a uniform bar of some suitable n-type material, with ohmic contacts. At present, GaAs and InP appear to be suitable practical materials. The devices made from these materials can show very high frequency oscillations when biased with strong enough fields and operated in circuits similar to that shown in Fig. 8.2. The oscillations are caused by electrons transferring from a high mobility state to a low mobility state under the action of strong electric fields. Such devices, discussed in section 8.5, are called *transferred electron oscillators*, or *Gunn diodes* after J. B. Gunn who discovered the effect experimentally.[8.15]

8.2   Energy balance and scattering limited velocities

It has been seen that at high electric fields, the electrons are not in thermal equilibrium with the atoms of the crystal lattice. The equilibrium distributions of energy amongst the electrons cannot be taken as any of the standard forms. In fact, complicated computer analyses are required to find such detailed information.[8.18,8.22] If rapid progress is to be made in the physical understanding of hot-carrier processes it is necessary to appeal to simpler arguments, even if of less precision. The principle of energy balance is one such useful argument, as will be seen in the paragraphs to follow.

Consider a gas of mobile electrons, moving around in a crystal. The thermal vibrations of the atoms about their mean positions prevent the crystal from being *perfectly* periodic. Electrons then 'collide' with the phonons associated with the lattice vibrations. The various ways in which electrons could lose energy to the crystal lattice were briefly outlined in section 6.7. Here, it is supposed that these losses of energy combine in such a way that an electron gas with an average energy $\mathscr{E}$ per electron loses energy to the crystal lattice at a rate $(\mathscr{E} - \mathscr{E}_0)/\tau_e$ where $\mathscr{E}_0$ is the average energy per electron if the electron gas were in equilibrium with the crystal lattice. The characteristic time τ_e is the *energy relaxation* time. The concept of such a time allows one to balance the rate at which electrons gain energy from an applied field against the loss of energy to the lattice. On average, the electrons gain energy at a rate eEv, where v is the average drift velocity in the field E. The energy balance equation is then

$$(\mathscr{E} - \mathscr{E}_0)/\tau_e = eEv \qquad\qquad 8.1$$

In some examples it is convenient to assign a 'temperature' to the electrons so $\mathscr{E} = \frac{1}{2}mv^2 = \frac{3}{2}kT_e$ with $\mathscr{E}_0 = \frac{3}{2}kT_0$. The 'temperature' T_e is known as the *electron temperature* though it must be appreciated that this does not imply a thermodynamic state because it does not involve a thermal equilibrium distribution of energies.

The relationship given by eqn 8.1 should be compared to the momentum balance derived earlier: $mv = eE\tau_m$ (eqn 3.3). This relates the average electron velocity to the field. The value of τ_m can be estimated from the mobility. In the simplest theories $\tau_m = \tau_e = \tau$, where τ is some free time between 'collisions'. However, there are many collision processes which hardly change the electrons' energy although they may change the momentum by a change of direction of the motion. Consequently the momentum gained by the electrons from the applied field, and usually in the direction of the field, can be lost rapidly while the collisions merely redistribute the energy; the energy is then lost at a slower rate than the momentum. Thus usually one has $\tau_m < \tau_e$.

From eqns 8.1 and 3.3, assuming constant values of τ_m and τ_e,† one can see that the electrons' excess energy increases as E^2. This shows that the energy lost, per interaction, to the crystal lattice will eventually reach the energy of the most energetic phonon associated with the crystal lattice vibrations. It was stated (section 6.7) that the periodicity of the crystal lattice imposes an upper frequency limit on the lattice vibrations; this in turn imposes an upper energy limit, $\mathscr{E}_{opt}$ on the phonons. The collision process may be pictured as an interaction between two particles. This indicates that the energy exchange per interaction will be limited to $\mathscr{E}_{opt}$. If the electrons energy is to be lost more rapidly, then the frequency for phonon–electron collisions must increase. If τ_{ph} gives the mean free time between phonon collisions, then the energy balance at high fields would give

$$\mathscr{E}_{opt} = eE\tau_{ph} \qquad\qquad 8.2$$

However it may be shown that energetic phonons also have considerable momentum, and one finds that τ_m and τ_{ph} must be closely linked. Putting $\tau_m \approx \tau_{ph}$ allows the field, E, to be eliminated from eqns 3.3 and 8.2 to give

$$mv^2 \approx \mathscr{E}_{opt} \qquad\qquad 8.3$$

From this argument it is expected that the velocity of charge carriers would be independent of the field—given high enough fields. Indeed if bars of Si or Ge, shaped as in Fig. 1.5, are subjected to high-enough voltages one finds that the current limits. The dumb-bell shape is used to avoid anomalous effects at the contacts. Experiments of this type gave the first indication that carrier velocities did indeed limit, to values around 10^5 m/s. More accurate methods of determining this saturated velocity are discussed elsewhere.[8.1] Figure 8.3 shows typical results for Si electrons. Similar results hold for holes and the carriers in Ge. The velocity field characteristic

† In general τ_m and τ_e must be regarded as functions of the electrons' energy.

for GaAs will show distinct differences from Fig. 8.3, as discussed under Gunn effect.

In many microwave devices, the transit time of the carriers determines the dimensions. A high value of v_s would be advantageous in most of the different types of transistor. It is believed that substantial increases can be made in this limiting value by the engineering of ternary alloys which form semiconductors. It has been suggested that $InAs_{0\cdot4}P_{0\cdot6}$ could be just such a useful material with a scattering limited velocity around $4\cdot10^5$ m/s.[8.21]

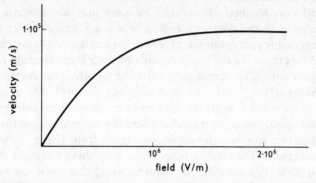

Fig. 8.3 Velocity saturation (data appropriate for electrons in Si; holes and electrons in Ge and Si behave similarly).

8.3   Avalanche breakdown: The Zener diode[8.5]

When a valence electron escapes from its parent atom to the conduction band, leaving behind a hole, the atom is said to be ionized. The process of ionization requires a *minimum* energy of $\mathscr{E}_g$, the band gap energy. The valence electron can receive this energy in a number of ways: thermal vibrations of the crystal (thermal ionization), energy from incident light (photoionization), or impact from a high-energy charge carrier (impact ionization). The incident charge carrier in an impact ionization has to have a slightly higher energy than the band-gap value because, in creating both a hole and electron by the ionization process, momentum as well as energy has to be conserved. It is then found that the incident charge carrier is unable to lose all its energy (Problem 8.1). There is also some threshold energy $\mathscr{E}_{th} > \mathscr{E}_g$ for the charge carriers before impact ionization can occur. With large enough electric fields, a few electrons can reach this energy. It is this impact ionization that leads to the process called *avalanche breakdown* that is of interest here.

An important parameter in the process is the *ionization coefficient* α for electrons (β for holes). In a distance δl, where α is constant, one electron will produce, by impact ionization, $\alpha\ \delta l$ more holes and $\alpha\ \delta l$ more conduction electrons. Thus α is defined by the probable number of ionizing collisions per unit distance of drift by the electron. The parameter β gives the analogous property for holes. (If the reader finds it difficult to conceive of a hole having a collision, he should remember that the hole is an electron wavepacket with energy and momentum. It is not merely an absence of an electron.) The parameters α and β are not in general equal, but a lot of the physics is more readily appreciated by putting $\alpha = \beta$. Now in a conductor one could find very large conduction currents flowing before the avalanche breakdown values of field were reached. It is therefore more practical to examine the ionization process in the depletion region of a p-n junction under reverse bias. Here negligible current flows, at least until breakdown is approached. It is assumed that there is a depletion region of length L and the field E varies over this region with a corresponding variation of $\alpha(E)$. An average value $\overline{\alpha L}$ is defined by $\overline{\alpha L} = \int_0^L \alpha\,\mathrm{d}x$, so that $\overline{\alpha L}$ gives the probable number of electrons or holes when a single charge carrier traverses the whole length of the depletion region. Suppose now that there is a current, I_{sn}, of thermally generated electrons entering the depletion region from the p$^+$ contact to the depletion region. The electric fields will sweep these electrons through the depletion zone, but for every electron crossing there will, on average, be $\overline{\alpha L}$ pairs of holes and electrons liberated into a mobile state. These holes and electrons will travel in opposite directions and so between them drift a total distance of L for each pair, no matter where the pair is produced initially. Because $\alpha = \beta$, each pair will, in turn, produce $\overline{\alpha L}$ more hole–electron pairs, and so for every electron that enters from the p$^+$ region there will be $(M - 1)$ more electrons and holes produced by an avalanche effect. M is the *multiplication factor* and it may be seen that

$$M = 1 + \overline{\alpha L}(1 + \overline{\alpha L}[1 + \overline{\alpha L}\{\cdots = 1 + \overline{\alpha L} + (\overline{\alpha L})^2 + (\overline{\alpha L})^3 + \cdots$$

$$= 1 \bigg/ \left[1 - \int_0^L \alpha\,\mathrm{d}x \right] \qquad\qquad 8.4$$

There is then a current MI_{sn} created by electrons reaching the n$^+$ contact to the depletion region, and a hole current $(M - 1)I_{sn}$ at the p$^+$-contact. Similarly, a hole current I_{sp} of thermally generated holes at the n$^+$-contact will generate a current MI_{sp} in an analogous manner. It should be particularly noted that, although the proportion of the current carried by the electrons and by the holes varies, the total current remains the same throughout the region at a value $M(I_{sn} + I_{sp})$. Figure 8.4 indicates how the holes multiply up from one side and the electrons from the other. If the ionization rates are not equal, then the multiplication factors for the injected holes and the injected electrons will differ, but qualitatively the process is similar.[8.1,8.3,8.4]

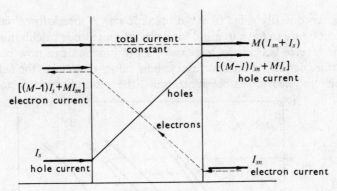

Fig. 8.4 Multiplication of electron and hole currents (note electrons and holes move in opposite directions but currents add).

It must be appreciated that the ionization rates change rapidly with the electric field. Typically one finds that

$$\alpha = \alpha_0 \exp\left[-b(E_0/E)^m\right] \qquad m \sim 1 \text{ to } 2 \qquad 8.5$$

Figure 8.5 shows the variation of the ionization rates for GaAs and Si. A rough appreciation of this form of dependence is obtained by considering the energy balances with the concept of a mean free path. If a charge carrier is to reach the threshold energy $\mathscr{E}_{th}$ required for ionization, then it must be free to travel a distance l such that $eEl = \mathscr{E}_{th}$. However, if the mean free path is $l_0 < l$, it follows that only a fraction of the electrons can gain the threshold energy. If $P(x)$ is the probability of a collision in which all the energy is lost after a distance x from the previous collision, then the probability of a collision between x and $x + dx$ is given by the joint probability $P(x)$ and dx/l_0 (dx/l_0 being the probability of a collision within any distance dx). The same probability must be given by $P(x) - P(x + dx)$; thus for small-enough dx

$$\partial P/\partial x = -P/l_0 \quad \text{or} \quad P(x) = \exp{-x/l_0} \qquad 8.6$$

The arbitrary constant for the solution to the differential equation has been fixed in eqn 8.6 by noting that a collision after $x = 0$ is certain (i.e. has unity probability). It follows that one will expect a fraction $\exp\left(-\mathscr{E}_{th}/eEl_0\right)$ of the electrons, at any position, to have sufficient energy to ionize an atom. The ionization rate would then be expected to be closely linked to this form. Evaluation of the ionization parameters in more detail requires more complex arguments that are left for further reading.[8.6,8.4] The above analysis is presented merely to make it plausible that the ionization rates vary rapidly with field. So rapid is this variation that one often refers to a 'breakdown' field. For example, in Si, once the fields reach $4 \cdot 10^7$ V/m, the sample has to be extraordinarily short to avoid breakdown. A 'breakdown' field is often a useful figure to have in mind for order-of-magnitude calculations.

The reverse-bias characteristics of a p-n junction can now be explained in further detail. At modest reverse-bias values of voltage, the fields are low enough to give negligible ionization—the multiplication factor is unity

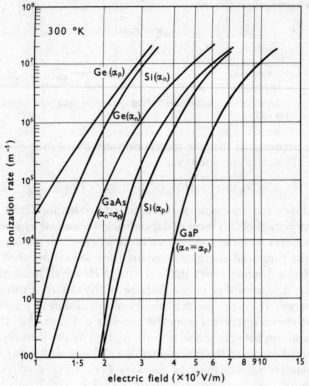

Fig. 8.5 Ionization rates for Si, GaAs, Ge and GAP (after S.M. Sze, reference 8.10, with permission).

and the reverse current takes the saturation value $I_s = (I_{sn} + I_{sp})$. At higher voltages, and hence higher fields, M becomes greater than unity, so that the reverse current becomes MI_s and, at a critical voltage, $M \to \infty$. The current is then controlled by the external circuit while the voltage changes little over a wide range of current values (Fig. 8.1). In Chapter 4 it was seen that the peak electric fields in p-n junctions were controlled by the impurity content at a given voltage. Thus the breakdown voltage has to be controlled by the choice of the impurity level. These diodes are commonly called *Zener diodes* and are used for voltage stabilizers or reference voltages. The true Zener effect, named after C. Zener, is one in which the electric fields reach the peak values (around 10^8 V/m) required for tunnelling of the valence electrons into the conduction band. Diodes showing this effect behave very similarly to those showing avalanche impact breakdown. In Si p-n diodes, the Zener effect occurs if the breakdown voltage is less than 5 volts approximately (see Problem 8.2). Reference 1.5 gives detailed values of breakdown voltage and required impurity levels.

Not all diodes show sharp breakdown characteristics. In some diodes the current slowly increases with reverse bias (soft breakdown). This is often caused by edge effects or defects of construction. Local dislocations of the crystal have also been found to give regions of *high* avalanche multiplication, forming dense regions of holes and electrons. These local regions are called *microplasmas* and show up as 'kinks' in the *I-V* characteristics close to breakdown.

8.4   The avalanche oscillator[8.8 – 8.13]

As noted in the introduction, p-n junctions under reverse bias conditions can oscillate on reaching avalanche breakdown, the oscillations occurring at very high frequencies. Typical outputs range from 100 mW at 50 GHz to 1 W at 10 GHz. Generically, oscillators of this type may be called Avalanche Oscillators. Acronyms such as ATTO (Avalanche Transit Time Oscillators) or IMPATT (IMPact Avalanche and Transit Time) have been developed to describe one type; another mechanism is described by the acronym TRAPATT (TRApped Plasma Avalanche and Triggered Transit). The basis for the modern Impatt diode was first described by W. T. Read in 1958.[8.8] He described the theory for a diode with four layers, n^+-p^+-i-p^+, the p^+ middle region forming a very thin avalanche zone. Current theories do not believe that the *very* thin avalanche region is essential and later developments of Read's theories show that simple p^+-n-n^+ (or p^+-p-n^+) diodes work well and in a manner close to Read's ideas. This section will briefly explain these current ideas.

Figure 8.6*a,b* indicates typical electric field profiles for p^+-n-n^+ Impatt diodes under reverse bias conditions where avalanche has started. The donor density is chosen so that the n-region is fully depleted (punched-through) before ionization occurs. Because of the rapid variation of the ionization rates with field, the avalanche multiplication occurs close to the peak field in a reasonably well-defined zone of length L_a (typically one-third of the n region as in Fig. 8.6*c*). This region is assumed to remain substantially constant in length even if the voltage across the diode is varied, though the rate at which the charge carriers are generated by ionization will vary as the field varies. Again it is assumed for simplicity that the field is large enough everywhere for the carriers to move at their scattering limited velocities. It is also assumed here that the electric fields (called space-charge fields) arising from the mobile charge carriers are negligible. The change in field across the avalanche zone is then proportional to the diode's voltage. The breakdown condition ($\int_0^{L_a} \alpha \, dx = 1$) corresponds to the steady state with a constant current. If the diode voltage moves above the value for this steady breakdown state then the current increases rapidly. The rate of

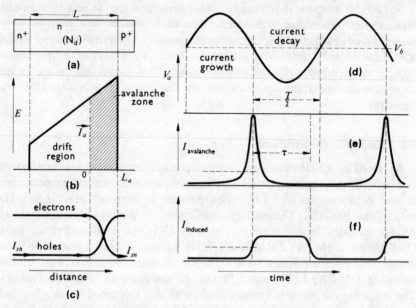

Fig. 8.6 The Impatt diode. (a) Structure. (b) Field profile: $dE/dx = eN_d/\varepsilon$, E everywhere large enough to induce scattering limited velocity. (c) Current multiplication in steady state showing avalanche zone. (d) Voltage variation with time, V_a being the voltage in shaded area of avalanche zone. (e) Current of electrons fed into drift region. (f) Induced current at terminals, transit time τ close to $\frac{1}{2}$ period. N.B. electron pulse drifts to left.

increase of the current is expected to be proportional to the excess of $\int_0^{L_a} \alpha \, dx$ over and above unity. Moreover, because the avalanche process is one of *multiplication*, the rate of increase of the current will also be proportional to the current already flowing through the avalanche zone. One would also expect the multiplication process to occur with a characteristic time constant dependent on the transit time of the charge carriers through the avalanche zone. W. T. Read put these physical ideas on a reasonable mathematical basis and developed an equation

$$dJ/dt = J\left\{\int_0^{L_a} \alpha \, dx - 1\right\}\Big/\left(\tfrac{1}{2}\tau_a\right) \qquad 8.7$$

where J is the current density through the avalanche zone and τ_a is the transit time of the charge carriers across the avalanche zone. Several simplifying assumptions are made, such as equal ionization rates and equal scattering limited velocities v_s for both the holes and electrons. Thus the transit time τ_a is given by L_a/v_s. Figure 8.6d,e then indicates relationships between a rapidly varying field across the avalanche zone and the current through the zone. So long as the voltage remains above the breakdown value the current grows, and only decays when the voltage (and hence field) fall below the breakdown condition. The extreme non-linearity of the current-generation process leads to the current achieving a very non-sinusoidal waveform with a sharp maximum (changeover between current growth and decay) as the voltage crosses the steady breakdown value. The peak in the current lags the peak in the voltage and this is characteristic of an inductance. An avalanche particle current is often termed inductive, though in general its equivalent inductance is highly non-linear (Problem 8.4 gives a linear estimate based on Read's theory).

Considerable simplification is obtained by approximating to the avalanche current with a pulse of charge: charge Q of holes emerging one side (only to disappear directly into the contact) and a charge $-Q$ of electrons to emerge the other side of the avalanche zone and to subsequently travel through the depletion region. The mean current is given by the charge Q per cycle, so that $I_0 = fQ$ where f is the frequency of the oscillation or disturbance. In the steady state the mean particle current I_0 through the avalanche zone is also the mean terminal current. However, as far as the instantaneous current is concerned, the time taken for the pulse of electrons to travel across the depletion region creates significant differences between the *induced* terminal current (Fig. 8.6f) and the avalanche current. This induced current is one of the useful methods for calculating the terminal current, and must now be discussed.

The induced current

If the current flowing into any element of a region bounded by two terminals (Fig. 8.7) is I, then it follows from the sum of both displacement and charge carrier current, that

$$I/A = J + \varepsilon_0\varepsilon_r \frac{\partial E}{\partial t} \qquad 8.8$$

The current I is the *total* current and it is this terminal current that is constant throughout the whole device (see Problem 3.3). It follows that it is not sufficient to know the rate at which charge carriers flow into the terminals of a circuit from the device; one must also know the displacement current in the device. However, when charge carriers are moving around, the changes of internal field from the average value can be very significant. The accurate determination of the displacement current thus appears to require the accurate determination of all the fields, but if eqn 8.9 is integrated over the whole length and one notes that $\int_0^L E \, dx = V$ (using the sign convention of Fig. 8.7), then eqn 8.9 gives

$$I = (A/L) \int_0^L J \, dx + C_0 \frac{\partial V}{\partial t} \qquad (C_0 = A\varepsilon/L)$$

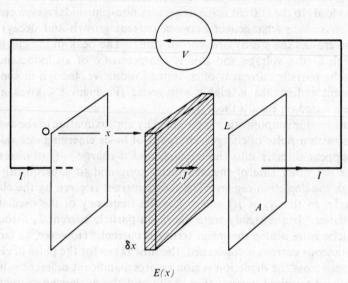

Fig. 8.7 Induced current. Block of charge induces a current equal to $J \, \delta x \, A/L$; capacitive current $C_0 \, dV/dt$, where $C_0 = A\varepsilon_0\varepsilon_r/L$, also flows.

The total current that flows into the terminals of the device is then composed of the capacitive current $C_0(\partial V/\partial t)$ and an *induced* current $(A/L) \int_0^L J \, dx$. This latter component is relatively easily calculated, as will be seen. The former component is the average displacement current across the region, so that C_0 is the normal 'parallel plate' capacitance between the bounding planes.

For the sharp-pulse approximation one should note that the pulse of holes travels directly into the p^+ contact (Fig. 8.6b) and hardly induces any current. The pulse of electrons, however, has to drift through the remainder of the depletion region to the n^+-contact. The electric fields in the depletion region are assumed to be high enough for these charge carriers to travel at their scattering limited velocity v_s. This is quite reasonable because the avalanche breakdown fields are more than an order of magnitude greater than the fields required to achieve velocity saturation in most semiconductors used for this device. If the pulse of electrons is narrow, its spatial distribution is of little concern and one can integrate the induced current as follows:

$$(A/L)\int_0^L J\,dx = (A/L)\int_0^L \rho v\,dx = v_s(A/L)\int_0^L \rho\,dx = Qv_s/L \qquad 8.10$$

Thus the induced current has a magnitude Qv_s/L for so long as the pulse is travelling across the depletion region, say for a transit time τ (not to be confused with τ_a). This then explains the roughly rectangular shape of the induced current component in Fig. 8.6f where, for the purposes of illustration, the period of the r.f. voltage is taken close to 2τ.

Negative resistance

Note should now be made that the capacitive current $C_0(\partial V/\partial t)$ is purely reactive and cannot contribute to the power flow. The induced current shown in Fig. 8.6f has, however, a fundamental Fourier component which is in antiphase to the avalanche field. If for the moment this field is assumed to be directly proportional to the applied voltage, then the terminal current is in antiphase to this applied voltage. The device then gives *out* power rather than absorbing it like a conventional positive resistance. Such a device is said to have a *negative resistance*. When it is placed in some suitable resonant circuit, with a resonance at the appropriate frequency, then oscillations in the current and voltage will spontaneously build up. The d.c. power will be partially converted into r.f. power which can be then fed into some load such as an aerial.

A very rough estimate of the efficiency can be made by noting that a good value of the transit time τ is around one-half of the period: $\omega\tau = \pi$. The induced current is then approximately a square wave with a fundamental Fourier component $4I_0/\pi$ where I_0 is the mean current. If the peak–peak r.f. voltage is $2V_1$ then the r.f. power developed is $\frac{1}{2}V_1(4I_0/\pi)$ while the d.c. input power is I_0V_0. The efficiency is then

$$\eta = (2V_1/\pi V_0)\cdot 100\% \qquad 8.11$$

The ratio of the r.f. to d.c. voltage is not given by the simple theory, but ratios around 1:5 fit more detailed theories quite well and also fit the experimental efficiencies—over 10 per cent for Si and 15 per cent for GaAs Impatts.

The range of frequencies where Impatt diodes give out power is not limited to the narrow conditions that $\omega\tau \sim \pi$. Typically a frequency range of 2 to 1 centred about the central frequency gives the order of the tuning available. The value for $v_s \sim 10^5$ m/s suggest that the 10 GHz diode will have a depletion width of 5 μm; higher-frequency devices will use much lower depletion widths. Laboratory devices have been made to give out powers up to several hundred GHz. Low-frequency Impatts are not often made because of the difficult task of getting the heat out from a wide depletion region and also because, at low frequencies, other oscillators (for example, using transistors) are better.

Heat dissipation is an important consideration in the construction of Impatts. The power density is so high that only small-volume devices with intimate contact to a heat sink can be made to work continuously (see Fig. 8.14). Problem 8.6 will give some idea of the magnitudes involved.

Microwave circuits

To obtain oscillations, a reverse-bias breakdown voltage with a steady d.c. current must be applied to the diode but simultaneously a r.f. voltage must be allowed to develop. Figure 8.8 shows a protype microwave circuit that illustrates some important points. The elements of the circuit are essentially the same for any two terminal negative resistance device, though the details will alter. Figure 8.8b shows the first-order equivalent circuit for the Impatt diode and its package, while Fig. 8.8c shows a typical package. The diode and its package behave as a negative conductance shunted by a

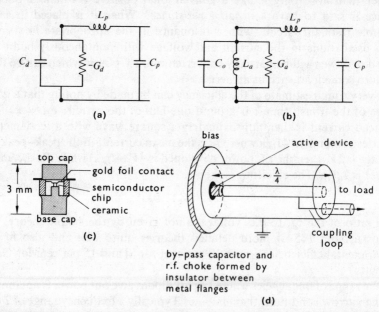

Fig. 8.8 Microwave circuit for active devices (schematic). (*a*) An equivalent circuit for Gunn diode and package. (*b*) An equivalent circuit for avalanche diode and package. (*c*) Package cross-section. (*d*) Features of microwave circuit.

reactance which is usually capacitive. The function of the circuit is to tune out this reactance at the required oscillation frequency. A coaxial line which is shorted at one end and has a length less than $\lambda/4$ (where λ is the oscillation wavelength) behaves as an inductance and can tune out a capacitive reactance. This provides the reason for the basic structure in Fig. 8.8d. The break in the coaxial circuit forms a block to d.c. but a by-pass capacitance to r.f. The loop coupling at the end of the coaxial line acts like a miniature transformer, coupling the external load into the microwave cavity and matching the negative conductance with an equal and opposite positive conductance at the oscillation frequency. Figure 8.2 indicates a simplified equivalent circuit for this microwave cavity. Practical circuits will have more complicated features, but the d.c. bias, tuning element and load coupling will always be present. Practical circuits will also have provision for varying the frequency. This can be done by changing the length of the circuit with a mechanical plunger, or with a varactor altering the capacitive reactance of the circuit.

Trapatts[8.11 – 8.13]

The electric fields that arise as a result of the charges generated by the avalanche have been ignored in the discussion so far, though they form important *practical* limitations. For Fig. 8.66, if E_1 is the field on the left-hand side of a pulse of avalanche charge $-Q$, and E_r is the field on the right, then $E_1 - E_r = Q/\varepsilon_0\varepsilon_r A$. Then the field near the avalanche is reduced below that shown while the field on the opposite side is increased. Consequently, if enough charge is generated sufficiently rapidly by an avalanche zone, then the field across the avalanche region will be lowered to such an extent that avalanche multiplication will cease. This provides one limitation on Impatt operation. Conversely, the electric field ahead of the pulse of charge is enhanced and, given sufficient charge, avalanche multiplication will start *ahead* of the original pulse and so provide an alternative limitation to the Impatt mode. These features also lead to a new mode of oscillation.

Suppose that voltage across the diode is raised very rapidly above the avalanche breakdown value. The multiplication and growth of the charge can be dramatic, producing a massive charge of both holes and electrons. Those charge carriers drifting across the depletion region raise the strength of the electric field ahead while decreasing it behind. The field ahead is forced into avalanche breakdown so that the avalanche zone moves ahead of the charge carriers (see Problem 8.7). The field behind falls to such a low value that the electrons and holes no longer move at the scattering limited velocity but regain their mobility. Then, as in any normal conductor with mobile charge, the charge distributes itself behind the avalanche zone in

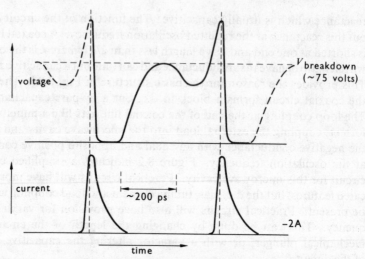

Fig. 8.9 Typical Trapatt waveforms.

an electrically neutral manner (see section 3.3 on dielectric relaxation). This neutral region contains both holes and electrons from the avalanche and is referred to as a *plasma*. The plasma is said to be 'trapped' because it lies in a low electric field. Thus the initial sharp rise in voltage across the diode, above its normal breakdown value, leads a moment later (a moment being 10–100 ps!) to a low-voltage and high-current state. Thus the key feature of the Trapatt mode is a sudden switching of the diode from a high-impedance high-voltage state into a low-impedance low-voltage state. Figure 8.9 shows typical waveforms that are observed across a Trapatt diode, though the discussion of circuits that are required to obtain this waveform must be left for further reading.[8.11] Efficiencies for the conversion of d.c. into r.f. are around 50 per cent for this mode of operation, though the frequency of operation is invariably lower than the corresponding Impatt mode using a diode of similar depletion width. The lower frequency is fundamental in that it takes longer to extract the trapped plasma than to remove charge carriers travelling at their scattering limited velocities as in the Impatt. Although currently showing great promise, the future of Trapatt oscillators depends on the solution of many technological difficulties—including the removal of heat, because even greater amounts are generated than in Impatt diodes!

8.5 The transferred-electron oscillator[8.14]

In 1963, J. B. Gunn discovered that oscillations could be obtained, at frequencies around 1 GHz, from short ($\sim 100 \ \mu$m) bars of n-type GaAs with ohmic contacts at either end.[8.15] The oscillations appeared in the current whenever the voltage exceeded some critical value around 0·3 volts per micrometre of material length. Research soon established that the mechanism was one of transferring the conduction electrons from a highly mobile state into a low mobility state. The energy for the transfer process was obtained from the electric field and could only occur when the field exceeded some critical value. This process had in fact been described previously in theoretical work but not proven until discovered by Gunn in his experiments.

The physical basis of the electron transfer can be outlined as follows. Firstly the material has to have a special band structure. Figure 8.10 shows one such suitable structure for the conduction band where a central valley in the $\mathscr{E}$-k relationship is surrounded, symmetrically, with satellite valleys which have a higher energy minimum. The upper energy valleys have a much lower mobility than the lower energy valley. At low electric fields the electrons satisfy a range of $\mathscr{E}$-k values in the lowest energy valley and have a correspondingly high mobility. As the electric field is increased, so the energy of the electrons increases, and at some critical field there is a probability of more electrons going into the upper valley than into the lower

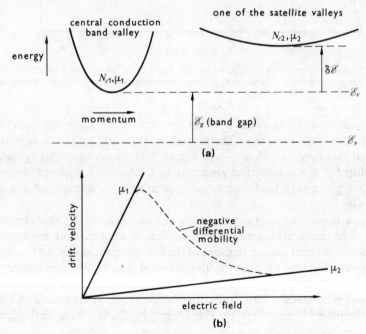

Fig. 8.10 Electron transfer. (a) $\mathscr{E}$-K characteristics for suitable material. (b) Velocity–field characteristics (schematic).

valley. This probability of transfer is greatly enhanced, and the critical field brought to within obtainable values, by the upper valley having a very low effective mass and so having an effective density of states N_{c2} that is much larger than the effective density of states N_{c1} for the lowest valley.†More advanced theory shows that this is also compatible with the upper valley having a much lower mobility than the lower valley. When the electrons transfer it follows that the average mobility will fall. Given

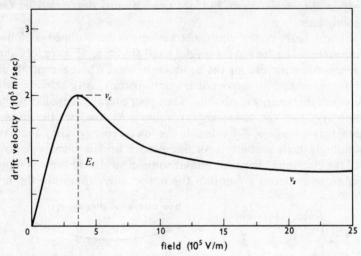

Fig. 8.11 Probable velocity-field characteristics for GaAs (data taken from Fawcett's calculation that is in best agreement with experimental measurements; reference 8.18, with permission).

suitable values of the change in mobility μ_1 to μ_2, the energy jump $\delta\mathscr{E}$, and the effective masses, then the *average velocity* of the electrons actually *decreases* as the *field increases*. The material is then said to have a negative differential mobility: $dv/dE = -\mu_d$. Figure 8.11 shows the velocity–field relationship for the conduction electrons in GaAs, with uniform electric fields. At high enough fields the velocity saturates in this material just as for Si or Ge.

There are important limitations on the use and validity of these characteristics. The relationships imply that the field is uniform and, curiously, we shall see in the next section that one of the first observations of the transfer effect showed that the field across the material was highly non-uniform. The response of the velocity to changes in field is also assumed to be immediate. However it takes a finite time for the electrons to transfer and this imposes frequency limitations on the effect. In GaAs, these limitations

† From eqn 3.21 it can be seen that $n_1/n_2 = (N_{c1}/N_{c2}) \exp \delta\mathscr{E}/kT_e$; consequently the effective electron temperature for transfer to occur is reduced by a factor of about $\log_e (N_{c2}/N_{c1})$.

are believed to make utilization of the effect difficult above 50 GHz; in InP they are possibly not so serious.

Domain formation

We now consider how the negative differential mobility of a material, as in Fig. 8.11, can give rise to a non-uniform field quite unlike that of a normal resistive material. From the discussion on dielectric relaxation (page 39) it follows that if dv/dE is negative, then any accumulations of charge will grow rather than decay. Indeed the growth rate can be very rapid. For example in GaAs, where the negative differential mobility is about one-third of the positive mobility at low fields, the characteristic time constant for growth will be around 3 picoseconds for n-type material with 10^{21} donors/m^3. A uniform distribution of charge is thus highly unstable. Any small accumulation spontaneously grows and any slight deficit of carriers moves towards becoming a depletion region. The initial experiments with GaAs showed this instability in the form of a triangular pattern of high electric field (Fig. 8.12) which nucleated, in general, at the cathode, moved through the sample at about 10^5 m/s, collapsed on reaching the anode, only to re-nucleate again at the cathode.

The reasons for this behaviour of the sample can be qualitatively understood by consideration of the velocity-field relationship. Consider then a sample of n-type GaAs of donor density n_0, length L and cross-sectional area A. At low voltages the field is uniform across the sample, as it would be across any uniform resistance. If the voltage is raised until $V_t = E_t L$ then the current reaches a peak value $I_t = e n_0 v_t(E_t) A$ and the mobility then starts to change from a positive differential value to a negative differential value. Some local region of material will transfer first of all (this region is usually found to be close to the cathode). The electrons in front and behind this region will move slightly faster than those in the middle. The faster ones from behind pile into the rear and form an accumulation of charge, while the front electrons tend to move away and initiate a charge deficit. These changes grow until a sharp accumulation layer and almost complete depletion region are formed (Fig. 8.12b). Gauss' theorem shows that this dipole layer of charge keeps the field high at the centre of the disturbance. The central electrons, and indeed the whole domain pattern, is found to move at a velocity close to v_s, the limiting velocity of the electrons in the material, as would be expected from the velocity-field characteristic. If the domain pattern is to remain stable and have a fixed voltage then one expects the accumulations and deficits of charge to remain unchanged, and in turn this leads one to suppose that the electrons outside the domain cannot be moving significantly faster or slower than the domain. If then a voltage

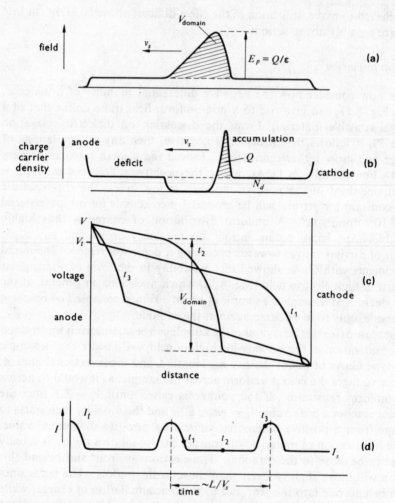

Fig. 8.12 (*a*) Field, showing 'dipole' domain. (*b*) Charge density creating domain. (*c*) Voltage variations with time. At t_1 domain is forming while old domain is leaving at anode; at t_2 domain is fully formed; at t_3 domain is leaving at anode and new domain has hardly formed at cathode. (*d*) Current variations with time.

$V > V_t$ is applied across the device, a domain must form to absorb some of this voltage and allow the field outside the domain, where the electrons will be moving close to the velocity v_s, to fall to E_s. Thus the current is reduced from its peak value I_t (occurring when the domain starts to form) to a value $I_s[\sim en_0v(E_s)A]$ corresponding to the velocity of the electrons outside the domain. As the domain moves, so the current I remains close to I_s until the domain runs into the anode. At the anode contact, the fields are no longer high enough to sustain the domain, which then disappears. Provided the voltage V is still in excess of V_t, the current will rise again to I_t and a new domain will be formed and travel through the sample. Consequently the current between I_s and I_t changes, in pulses, as a domain leaves at the anode and is re-formed (Fig. 8.12d). The pulse repetition frequency is $(v_s/L) \sim (10^5/L)$ Hz approximately. This summarizes the main features that were initially observed in Gunn's experiments.

Subsequent work showed that domain formation was not always essential, and that the field could redistribute itself over the sample in other ways (e.g. a high field towards the anode) and still have oscillations in suitable circuits. For domains to form as above, it is usually considered that a product of donor density × length, $[n \, . \, L]$, which is greater than $5 \, . \, 10^{15}$ donors/m^2 is required for GaAs—this allows the domain to grow in one transit. However, the detailed domain dynamics must be left for further reading;[8.1,8.14] we turn instead to consider practical oscillators which utilize this basic effect.

Gunn oscillators

Practical oscillators, using the transferred electron effect to give a few hundred milliwatts of power at 10 GHz, use epitaxial layers about 10 μm thick, with around 10^{21} donors/m^3, grown on n$^+$-substrates and possibly with another n$^+$ contact on the other side. Such devices will have areas around 10^{-8} m^2 and will be packaged in protective ceramic packages (see Figs. 8.8c and 8.14) and mounted in circuits similar to that in Fig. 8.8d. Alternatively they are bonded directly in chip form into a microstrip circuit laid out on thin sheets of alumina or other dielectric. It is believed that in these high-frequency devices the domain dynamics is distinctly different from that in low-frequency devices—probably the domain is dominated by an accumulation of charge rather than an equal accumulation and deficit of the triangular domain. The first-order equivalent circuit for these devices can be usefully taken as a negative conductance shunting a capacitance, though the values are best derived from experiment for accurate circuit designs. Although in Fig. 8.8d the circuit is shown to be the same for both Impatt and Gunn diodes it must be emphasized that there are distinct differences in the details of the values; moreover Gunn diodes usually have their d.c. power supplied at constant voltage while Impatts have their d.c. power supplied from a constant current source. Indeed, simple circuits like the one in Fig. 8.8d will not always give the best efficiency. It is generally accepted that a non-sinusoidal waveform can give better efficiencies for the conversion of d.c. into r.f. In Fig. 8.13 is illustrated, as one example, the ideas for a 'square' wave of voltage. To make the calculation simple one assumes that the domain's transit time is one-half of the period of the oscillation. A slight asymmetry in the waveform is required to ensure that a new domain is not triggered until the correct point in time. However the current changes roughly from I_t to I_s while the voltage swings from V_t to some substantial value V_s above V_t, with the domain absorbing the excess voltage. Both the current and voltage are approximately square waves. The power delivered at the *fundamental* frequency is then $\frac{1}{2}(2/\pi)^2(V_s - V_t)(I_t - I_s)$. The d.c. power input is $(I_t + I_s)(V_t + V_s)/4$, giving an efficiency for conversion of d.c. into r.f. of

$$\eta = \eta_m \frac{(1 - i)(v - 1)}{(1 + i)(v + 1)} \% \qquad 8.12$$

where $\eta_m = 8/\pi^2$, $v = V_s/V_t$ and $i = I_s/I_t$. The value of v is in principle only

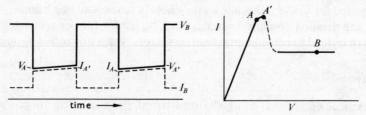

Fig. 8.13 'Square' wave operation. Right-hand diagram shows assumed terminal current–voltage characteristics.

limited by the bias available, though breakdown fields also limit the permitted bias. The value of i is determined by the material and the ratio of its peak velocity v_t to its valley velocity v_s (v_t/v_s is called the *peak-to-valley* ratio and a high value is essential for a high efficiency: this ratio for GaAs is about 2·5:1. InP is also believed to have a better value). The efficiencies given by eqn 8.12 are usually optimistic; with practical c.w. oscillators the efficiencies are typically less than 10 per cent. Pulsed oscillators at lower frequencies fairly regularly exceed 15 per cent and even 30 per cent has been claimed in one special example.

Practical devices can be made to have an octave of tuning range, or even more. The classical reasoning for this behaviour has always been considered to be a large r.f. voltage swing controlling the initiation and even the extinction of a domain. Recent work also suggests that the fields can be distributed in such a way as to give negative resistance across the terminals without domains cycling, although the material has $n \cdot L$ products such that domains should form. Contacts to the material also appear to have profound influences on the performance. Thus there are still many problems and areas where the theory is lacking. At present, transferred electron oscillators appear to work best around 10–20 GHz, with efficiencies falling off above 30 GHz. Devices made from InP appear to be promising, but at the time of writing their commercial development is not well established (see also next section).

Beside the obvious use for oscillators, the Gunn diode gives promise of applications to very high speed logic. The presence or absence of a domain can be detected and domains can be initiated, recycled and made to perform various logic functions.[8.14,8.20] At present these ideas are in the research stage.

The LSA oscillator[8.16]

If high-field domains could be prevented from forming so that the electric fields remained uniform in a sample, then the current–voltage relationships for such a device would be given directly from the velocity–field characteristics through $I = en_0 v(E)A$ and $V = E \cdot L$. The negative differential mobility would directly cause a negative differential resistance across the terminals of the device, and it has already been seen how a negative resistance can lead to oscillation. This mode of operation can be achieved and is known as the Limited Space-charge Accumulation mode (LSA).

The first practical method for inhibiting the growth of a domain, and so keeping a uniform field, was achieved by having an r.f. voltage that swept the field rapidly through the region of negative differential mobility. For a sinusoidal r.f. voltage, the amplitude, bias value and frequency had to be carefully controlled so that accumulations of charge were limited by the short time for which the sample was biased in the negative differential mobility region. Any accumulations of charge that might form were dispersed by an equally carefully controlled time in the positive differential mobility region. It can again be seen that a non-sinusoidal waveform can be very advantageous for this mode. For example, if the voltage changes from just below V_t up to a peak value V_s that lies in the saturated velocity range, and the change is made in a square wave manner, then the time spent in the negative differential mobility region can be negligible (at least in principle). The current would change from about I_t down to a value I_s, in a manner similar to that for a domain mode, but now there would be no restriction on the sample length or domain transit time. Long samples with high voltages could be used. Practical LSA oscillators tend to use simpler circuits that those required to approximate to a square wave of current and voltage. If there is an inductance in series with the Gunn diode then this inductance forces the voltage to increase rapidly as the current through the device falls. With the correct choice of load and inductance, dipole domain growth can be inhibited and LSA conditions prevail.

The key point about the LSA mode lies in its potential for high r.f. power outputs at high frequencies. Most semiconductor devices have relatively low peak voltages V_p, which in turn makes them require a relatively low resistance $(\sim V_p{}^2/2P)$ when operating at a high power level P. At high enough levels of power the resistive losses of the circuit become comparable to the optimum load, and then power is lost in the circuit rather than the useful load. The useful efficiency is reduced. Now a long sample of material operating in the LSA mode could have a much higher voltage then a device operating in a Gunn domain mode where the peak voltage is determined by the breakdown field in a relatively short domain. Much higher levels of

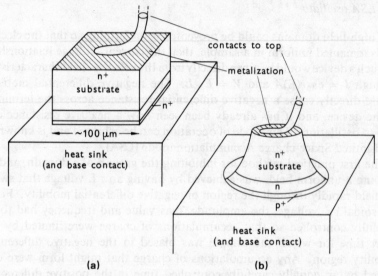

Fig. 8.14 Construction of Gunn (*a*) and avalanche (*b*) diodes (schematic). Both devices are typically bonded to heat sink by thermocompression bonding of gold layers on semiconductor and heat sink. The avalanche diode has a tapered shape made by etching; it is found to give protection against breakdown occurring at the edge rather than in the bulk. For clarity, epilayers are exaggerated in thickness.

power should then be obtained in the LSA mode before the low-impedance limitations occur. The LSA mode is then said to have a high *power–impedance* product. Indeed kilowatts of power have been obtained at frequencies below 10 GHz from samples several hundred micrometres in length.[8.17] However there are severe problems of removing the heat from such long samples. At shorter lengths, and usually higher frequencies (around 30 GHz), the problems of removing the heat are alleviated, but then at these higher frequencies the conversion efficiency has so far been found to be relatively low—around a few per cent for GaAs.

An alternative method of achieving LSA operation has recently been emphasized. In some materials the diffusion constant is artificially very high because the electron-transfer process creates a larger spread of velocities amongst the electrons than if they remained in a single valley. Diffusion tends to disperse charge accumulations and so inhibits domain formation. This is believed to be the situation in InP,[8.19] where possibly there are three types of conduction minima involved in the electron-transfer process. Certainly domain formation appears to be different to that in GaAs. Devices made from InP, operating at frequencies around 30 GHz, have shown considerable promise with substantially higher efficiencies than for GaAs—but their study is in an early stage.

In summary, it has been seen how energetic electrons create entirely new effects in the conduction properties of semiconductors. Some of these effects are very useful. However, to make successful hot-electron devices great strides have had to be made in the conduction of heat away from the semiconductor so as to keep the material relatively cool. Figure 8.14 gives schematically typical structures for Gunn and Impatt diodes, showing how the portion that generates the heat is placed close to the heat sink which is usually of copper—though in some cases pure diamond, with its better thermal conductivity, has been used! Thus, as always, good technology is the key to success—without it the devices become mere academic toys.

PROBLEMS

8.1 An electron with a kinetic energy E_k hits an atom and releases a hole and an electron with the loss of an energy E_g. If the three particles all have the same effective mass, show that the minimum E_k is $3E_g/2$ and that the final kinetic energy of the incident electron is one-ninth of its initial kinetic energy. Both momentum and energy are conserved in the collision process.

8.2 For estimating simple breakdown voltages in Si it is useful to use an effective ionization rate that is equal for both holes and electrons. A rough guide that compromises the values of breakdown over a range of peak fields of $10^7 – 10^8$ V/m is given by $\alpha = 10^5 (E/2.5 \times 10^7)^5$ m^{-1}. This gives the breakdown voltage to about 20 per cent accuracy even at the higher fields and shorter diode lengths. Consider then an abrupt one-sided p$^+$-n junction of depletion width d with $d = (2\varepsilon_0\varepsilon_r V_b/eN_a)^{1/2}$. Show that the breakdown voltage and peak field are related by $V = 3 (2.5)^5 \, 10^{30}/E_p{}^4$ where E_p is the peak field and the field changes from E_p to zero in a depletion distance $d = 60 (2.5 \, . \, 10^7/E_p)^5$ micrometres in this approximation. For electron tunnelling from the valence band into the conduction band the field has to be sufficiently strong so as to bring the conduction band in line with the valence band, but with only a spatial separation of around 10 nm. Show that this requires $E_p \sim 10^8$ V/m and hence show that a 3-volt breakdown diode would be expected to show Zener rather than impact ionization breakdown. See reference 1.5 for more accurate calculations.

8.3 The multiplication factor in the base-collector region of a transistor is $M \simeq 1/[1 - (V/V_b)^6]$. Show that the effective h_{fb} is Mh_{fbo} where h_{fbo} is the value at low enough fields for $M \sim 1$. Hence show that the breakdown V_{cbo} (when the emitter current controls the collector current) is given by V_b. Show also that $h_{fe} = Mh_{fbo}/(1 - Mh_{fbo})$ and hence the breakdown voltage V_{ceo} (when the base controls the collector current) is approximately $0.5V_b$ given $h_{fbo} = 0.984$.

8.4 Use the expression of eqn 8.7 with the value of the effective ionization rate given in example 8.2 to show that $j\omega J_1 = J_0 \times 10E_1/\tau_a E_a$ for small changes of current and field, J_1 and E_1, about the steady values J_0 and E_a. Hence show that the avalanche zone appears for small signals as an inductance of value $L = (\tau_a V_a/10I_0)$ where $I_0 = J_0 \times$ area and $V_a = E_a \times d$ for an avalanche of width d.

Show that one would expect the parallel plate capacitance of the avalanche zone to resonate with this inductance (the combined avalanche and capacitive currents) at a frequency given by

$\omega_a{}^2 = 10J_0/\tau_a E_a \varepsilon_r \varepsilon_0$. This is called the avalanche resonance frequency ω_a. More detailed theory shows that one typically requires to operate above the avalanche resonance frequency.

8.5　With an avalanche resonance frequency taken as 10 GHz and the avalanche zone as 1 μm wide, estimate the current density that is required. For a device 100 μm square, show that over 5 watt goes into the avalanche zone itself, making no allowances for the dissipation in the drift space of the avalanche diode. (Take $\varepsilon_0 \varepsilon_r \simeq 10^{-10}$ F/m and $F_b \simeq 4 \times 10^7$ V/m).

8.6　An idealized Gunn domain has diffusion neglected so that the accumulation region is assumed to be very narrow. Show that the sample can absorb a voltage $[E_s L + \frac{1}{2}(en_0 d^2/\varepsilon_0 \varepsilon_r)]$ where L is the length of the sample with a donor density n_0 and depletion distance in the domain of d. The field outside the domain is E_s. Show also that the energy stored in the domain is $\frac{1}{6}\varepsilon_0 \varepsilon_r E_p{}^2 d$ per unit area where E_p is the peak field ($E_s \ll E_p$). Show that if the peak field is 3·2 $\times$ 10^7 V/m then the domain width is approximately 10 μm with a domain voltage of about 160 V (n_0 taken as 2 . 10^{21}/m^3). If the sample is 30 μm long and the domain cycles at a velocity of 10^5 m/s show that a power of over 5 W, for an sample 100 μm square, could be delivered from the domain energy *if* it could be transferred directly into the circuit.

The assumption of zero diffusion is in general very poor. Discuss the qualitative effects that you would expect diffusion to have (see reference 8.1).

8.7　In an asymmetric diode with a donor density n_0 for the lightly doped region, the breakdown is assumed to occur sharply at the field E_b. Show that if a current density $en_0 v_0$ is fed into the terminals of the device then the point at which the field reaches the value E_b will travel at v_0 provided that v_0 is greater than the saturated velocity of the charge carriers. (*Hint*: note that before any avalanche multiplication occurs the diode is depleted so that the displacement current is the only current; also use $\partial E/\partial t = v_0 \, \partial E/\partial x$. In region where avalanche has occurred the current is assumed to adjust itself automatically). Hence show that the avalanche zone may be expected to propagate through the diode.

8.8　Inversion of the n-type semiconductor at a suitable metal–semiconductor interface was discussed with reference to Fig. 7.1e. Show that a similar effect could happen at a p$^+$-n interface that was appropriately abrupt. Show that the electric field would be expected to give an additional peak close to the p$^+$-n junction. How would this affect the breakdown of the p-n junction on reverse bias?

8.9　Consider the power that can be given out from a negative conductance $-G_n$ as compared to the power absorbed by a positive conductance G_p. Hence show that, if there is a shunt resonant circuit across the parallel combination of the positive and negative conductance, oscillations will grow at the resonant frequency provided that $G_n \geqslant G_p$. If G_p is formed from the properly terminated transmission line of characteristic impedance $1/G_p$ show that the voltage reflection coefficient ρ is given by $\rho = (G_p + G_n)/(G_p - G_n)$ at resonance, provided that the device is not oscillating. Hence explain how to use a microwave circulator to form an amplifier of maximum gain $|\rho|^2$. Explain why, in Impatt and other negative resistance devices used as amplifiers, the power gain is usually limited to less than 15 dB (Hint consider $\partial \rho/\partial G_n$).

General references

SZE, S. M. Reference 1.5.
8.1　CARROLL. J. E. *Hot Electron Microwave Generators*. Edward Arnold, 1970.
8.2　WATSON, H. A. *Microwave semiconductor devices and their circuit applications*. McGraw-Hill, 1969.
8.3　MOLL, J. L. *Physics of Semiconductors*. McGraw-Hill, 1964.

Special references

Avalanche breakdown and Zener diodes

8.4　CHYNOWETH, A. G. Charge multiplication phenomena. In *Physics of iii–v compounds*, ed. R. K. Willardson and A. C. Beer. Academic Press, 1967.
8.5　TODD, C. D. *Zener and Avalanche Diodes*. Wiley, 1970.

8.6 BARAFF, G. A. Distribution functions and ionization rates for hot electrons in semiconductors. *Phys. Rev.*, **128** (1962), 2507–17.

8.7 SZE, S. M. and GIBBONS, G. Avalanche breakdown voltages of abrupt and linearly graded p-n junctions in Ge Si GaAs and GaP. *Appl. Phys. Letters*, **8** (1966), 111–2.

Avalanche oscillator diodes

8.8 READ, W. T. A proposed high frequency negative resistance diode. *Bell Syst. Tech. J.* **37** (1958), 401–446.

8.9 MISAWA, T. Impatt diodes. Chapter 7. In *Semiconductors and Semimetals*, Vol. 7, ed. R. K. Willardson and A. C. Beer. Academic Press, 1971.

8.10 SZE, S. M. and RYDER, R. M. Microwave avalanche diodes. *Proc. IEEE*, **58** (1971), 1140–54.

8.11 EVANS, W. J. Circuits for high-efficiency avalanche diode oscillators. *IEEE Trans.*, **MTT-17** (1969), 1060–67.

8.12 CLORFEINE, A. S., IKOLA, R. and NAPOLI, L. S. A theory for the high-efficiency mode of oscillation in avalanche diodes. *RCA Rev.*, **30** (1969), 397–421.

8.13 SLAYMAKER, N. A. and CARROLL, J. E. Verification of a simple Trapatt-oscillator model. *Proc. IEE*, **119** (1972), 1113–18.

Transferred electron devices

8.14 BULMAN, P. J., HOBSON, G. S., and TAYLOR, B. *Transferred Electron Devices*. Academic Press, 1972.

8.15 GUNN, J. B. Microwave oscillations of current in III–V semiconductors. *IBM J. Res. Devlpmt*, **8** (1964), 141–9.

8.16 COPELAND, J. A. *Semiconductors and Semimetals*. Vol 7, edited by R. K. Willardson and A. C. Berr. Academic Press, 1971.

8.17 EASTMAN, L. F. *Optimalization of High Power Pulsed LSA Oscillators, 8th Int. Conf. on Microwave and Optical Generation and Amplification*, Kluwer Deventer, Netherlands, 1970.
EASTMAN, L. F. *Gallium Arsenide Microwave Bulk and Transit Time Devices*. Artech House, Massachusetts, 1972.

8.18 FAWCETT, W., BOARDMAN, A. D. and SWAIN, S. Monte Carlo determination of electron transport properties in Gallium Arsenide. *J. Phys. Chem. Solids*, **31** (1970), 1963–90.

8.19 HILSUM, C. and REES, H. D. Three level oscillator: a new form of transferred electron device. *Electronic Letters*, **6** (1970), 277–8.

8.20 HARTNAGEL, H. L. Theory of Gunn effect logic. *Solid St. Electron.* **12** (1969), 19–30.

Other references

8.21 FAWCETT, W., HILSUM, C. and REES, H. D. Optimum semiconductor for microwave devices. *Electronic Letters* **5** (1969), 313–14.

8.22 CONWELL, E. M. *High field transport in semiconductors*. Academic Press, 1967.

9

Optical Devices

9.1 The photon

This chapter is concerned mainly with p-n junctions that either produce light or detect the presence of light. The underlying purpose of all the discussions is to illustrate the wide range of interactions that are possible between photons and electrons. Photons are the quantum mechanical particles of energy hf associated with electromagnetic radiation of frequency $f (= \omega/2\pi)$.† The properties of photons may not be as familiar, to most readers, as those of electrons, and so the chapter is started with this introductory section.

Electromagnetic radiation is associated with a combination of electric and magnetic fields that can be characterized by a wave function which varies as $\exp j(\omega t - kz)$. The phase velocity $c (= \omega/k)$ is the velocity of light for the medium, and in general the group velocity $c_g (= d\omega/dk)$ will be different from c. In such a dispersive medium the photons travel at c_g and their energy and momentum are related by $\mathscr{E} = pc$, though $\delta\mathscr{E} = \delta p c_g$. Figure 9.1 shows the relationship between the free-space wavelength and the photon's energy in electron volts. The visible optical ranges are emphasized on account of their special interest in this chapter. The large magnitude of c implies that photons have a very much larger wavelength than that associated with electrons of a similar energy. Another major difference arises because the phase of a wave of photons can be observed directly, for example by measuring the electric field in the wave. The phase of an electron wave can never be observed in the same sense. This difference is only made clear in more advanced quantum mechanical work[9.15] but in outline there is an uncertainty between any knowledge of the phase ϕ of a wave of photons and the number of photons N: $\Delta N . \Delta\phi \sim 1$. The phase of a single photon is not measurable. However, unlike electrons, photons do not obey the Pauli exclusion principle; one can have as many photons in the same energy and wave vector as one can generate. The phase can then be meaningfully observed. These basic differences are perhaps the reason

† Many books prefer ν for an optical frequency, though to avoid a change of notation we have here kept f for frequency.

why it is often easier to recognize the wave nature of photons as opposed to those of electrons.

The absence of any restriction on the number of photons in a given state confers on photons an entirely different statistical behaviour from that of electrons. The probability of finding a photon in a state of energy $\mathscr{E}$ is then given by the Bose-Einstein distribution

$$B(\mathscr{E}) = [\exp (\mathscr{E}/kT) - 1]^{-1} \qquad 9.1$$

Particles which obey this distribution are *bosons*. The derivation of

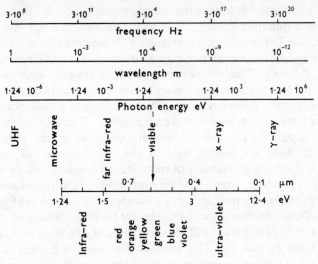

Fig. 9.1 Photon energy–wavelength–frequency.

this distribution is left, like the Fermi-Dirac distribution, for further reading.[3.3 – 3.7] The result however is used to find the equilibrium distribution of photons within an enclosed box of volume V. The discussion follows that given for electrons where, on page 140, it can be seen that a quantum state corresponding to a wave function $\psi \exp j(\omega t - \mathbf{k} \cdot \mathbf{x})$ requires a volume V_k in k-space such that $V_k \cdot V = (2\pi)^3$. For photons in a range of frequencies f to $f + \mathrm{d}f$ one requires a range of k vectors with magnitude k to $k + \mathrm{d}k$ where f and k are related by the expressions for group and phase velocity. Thus such a range of frequencies occupies a shell with a k-space volume given by $\delta V_k = 4\pi k^2 \, \mathrm{d}k = 4\pi \cdot 8\pi^3 (f^2 \, \mathrm{d}f/c^2 c_g)$. There should then be $\delta V_k/V_k$ photon states available in this energy range. However it may be shown that electromagnetic plane waves can have their electric field in any orientation (*polarization*) normal to the direction of propagation. This means that there are two independent polarizations for each k-vector state. Thus there is a factor of two, like the factor arising for the spin in electron waves. One can then multiply the number of states per unit volume by the probability of occupation $B_e(\mathscr{E})$ and also by the energy $\mathscr{E} = hf$ to arrive at the energy *density* $P(f)\, df$ for photons between f and $f + df$:

$$P(f) = 2(\delta V_k/V_k) \cdot (hf) \cdot B_e(hf) = hf \left(\frac{8\pi f^2}{c^2 c_g} \right) \left[\exp \frac{hf}{kT} - 1 \right]^{-1} \qquad 9.2$$

This relationship gives Planck's law of radiation. It must be emphasized that this is a distribution which results from thermal equilibrium and will, in general, be found only in equilibrium situations.

Of particular interest here is the interpretation of eqn 9.2 that is possible with modern quantum theory. It is supposed that the box is made of a semiconductor [or indeed any material] with electrons at energies $\mathscr{E}_1$ and $\mathscr{E}_2$ continuously interchanging energy with the photons of frequency f such that $hf = \mathscr{E}_2 - \mathscr{E}_1$. There is, of course, a whole range of such interactions that are possible, but it is only necessary to concentrate on two to demonstrate the required points. Some electrons at the higher energy $\mathscr{E}_2$ will spontaneously drop to the lower energy level $\mathscr{E}_1$ and emit photons. This is called *spontaneous emission*. It can only occur if there are electrons at the upper energy and also if there are vacant sites at the lower energy to receive the electrons.† Thus the probability of spontaneous emission can be written as $AF_2(1 - F_1)$/unit time, where F_2 is the probability of electrons existing at $\mathscr{E}_2$ and $(1 - F_1)$ being the probability of a vacancy at $\mathscr{E}_2$. Function F is given by the Fermi-Dirac distribution with the relevant Fermi level inserted. However the density $P(f)$ of radiation will also give some of its energy into raising the electrons from $\mathscr{E}_1$ into $\mathscr{E}_2$. This *stimulated absorption* process is expected to occur, using similar arguments to those above, at a rate proportional to $P(f)$ and given by $BP(f)F_1(1 - F_2)$, where B has yet to be related to A. Now in quantum mechanics it can be shown that every transition has a converse transition with an equal transition probability. Thus there is an induced or *stimulated emission* process whereby the radiation stimulates the electrons into losing energy. Stimulated emission is in addition to spontaneous emission, and is expected to occur at a rate $BP(f)F_2(1 - F_1)$; note the interchange of suffices 1 and 2 from the absorption case, the constant B remaining the same. The total rate of change of power density is thus

$$\frac{dP(f)}{dt} = hf\{AF_2(1 - F_1) - BP(f)[F_1(1 - F_2) - F_2(1 - F_1)]\} \qquad 9.3$$

In equilibrium $P(f)$ is constant so that

$$P(f) = (A/B)/\{[(F_2^{-1} - 1)/(F_1^{-1} - 1)] - 1\} \qquad 9.4$$

† c.f. recombination (p. 51) where it was seen that the probability of a transition was proportional to the product of the probabilities of the components taking part.

However, from Fermi-Dirac it follows that $F^{-1} - 1 = \exp{(\mathscr{E} - \mathscr{E}_f)/kT}$, so that eqn 9.4 is identical to eqn 9.2 provided

$$A/B = 8\pi h f^3/c^2 c_g \qquad 9.5$$

and $\mathscr{E}_2 - \mathscr{E}_1 = hf$. The coefficients A and B are called the Einstein coefficients of spontaneous and stimulated emission.

At low frequencies, it is not usual to think in terms of absorption or stimulation of emission because the discussion can be phrased in classical terms. For example the *linear accelerator*, used for nuclear studies, allows microwave electromagnetic radiation to travel along with electrons. The electrons are bunched into regions where the electric field in the wave accelerates them and gives them more energy. The electrons can then be said to be absorbing photons. The converse process occurs in the generation of microwave frequencies in a *travelling wave tube*. The electric wave is slowed down to travel with the electrons but now the electron bunches occur at the phase of the electric field, which decelerates the electrons so that they lose energy to the electric field—one could say that the electrons emit photons. At higher frequencies, spontaneous emission becomes far more probable than stimulated emission in most cases (Problem 9.1) and it is spontaneous emission that produces most of the illumination in our world.

As a final note in this section, photons tend to interact singly with electrons in solids, so that at each interaction there is an energy exchange of $\hbar\omega$, rather than multiples of $\hbar\omega$. Advanced quantum mechanics shows that multiple interactions require non-linearities in the medium or non-uniform fields over the interaction region. These non-uniformities are often weak and are not considered here. The simple mechanistic view, encouraged by the particle language, suggests that simultaneous interactions between three or more particles are less likely than those involving only two.

9.2 Optical interactions

The range of interactions between photons and solids is vast, and fills many books with detailed results and experiments.[9.1,9.2,9.3] Despite the complexity of the subject the practical results have been impressive: television, films, solar cells to power satellites, xerography for rapid reproduction of print, and lasers—to name but a few of the successes. However, only a few of the principles behind those optoelectronic devices that specifically use semiconductors will be considered here.

Figure 9.2 shows schematically some of the methods by which light can be absorbed in a semiconductor as a result of the electrons or holes moving to higher energy states. At low frequencies where $\hbar\omega$ is considerably less

208 OPTICAL DEVICES

than the band gap energy, the principle source of absorption comes from the free carriers moving to higher energies. In pure semiconductors this interaction is weak because there are few free carriers. Free carrier absorption will continue over a range of energies and, as the photon energy increases, electrons may be excited from impurity states into the conduction band, or holes into the valence band. As the photon energy approaches that of the band gap so excitons can start to absorb light. The exciton, as briefly

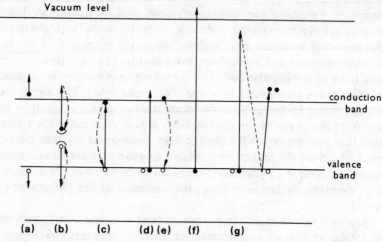

Fig. 9.2 Schematic diagram of absorption–emission processes. (*a*) Free-carrier absorption of photon energy with free carrier gaining energy. (*b*) Absorption or emission of photons with electrons or holes changing from impurity states to allowed bands. (*c*) Exciton absorption or emission of photons. (*d*) Absorption of photon with electron from valence band going to conduction band and leaving behind a hole. (*e*) Emission of photon by recombination of electron and hole across band gap. (*f*) Photoemission of electron into vacuum level. (*g*) Multiple hole–electron production by high-energy photon absorption.

discussed on page 138, is a hole–electron pair that behaves in many ways like the proton–electron pair of the hydrogen atom. There are a series of energy levels with separations below the band-gap energy, with the band-gap energy representing the energy at which the electron becomes free from the binding influence of the hole. At room temperature it is usually possible to detect only the principle absorption energy because of the very small separation of the discrete exciton levels. One must not take the 'hydrogen atom' picture of the exciton too literally; it is only one way of looking at the hole–electron coupling but the more advanced theories must be left for the specialist. For photon energies very close to the band gap, the light can be absorbed by transitions of the valence electrons into the donor impurity states. These states can form tails of states just below the conduction band, when they have a high density. These tails of impurity states lead to absorption of the light below the band-gap energy, though the absorption decreases exponentially with the decrease in energy below the band-gap value. For photon energies above the band-gap value the absorption probability increases sharply as the valence electrons are excited directly into the conduction band. This enhances the conductivity of the material—the effect called *photoconductivity*. At even higher energies the electrons can be excited into the vacuum level, i.e. escape from the solid. This is the principle of the photocathode, a device that emits electrons when suitable radiation shines on it. When the photon energy is much larger than the band gap (X-rays for example) the reaction is more complicated. The photons can excite electrons into the conduction band from the valence band and these electrons can still have sufficient energy to either escape from the solid or to ionize yet more atoms into producing hole–electron pairs. In general, only a few electrons escape from the solid because the energetic photons are absorbed deep inside the solid. Electrons liberated from the valence band then lose so much energy by collisions on their way to the surface that they cannot finally escape from the surface. Secondary X-ray emission may be produced by really energetic photons exciting the tightly bound electrons at fixed atoms, but these X-rays may also be absorbed if the volume of material is large enough. As a result of all these processes one finds that, in a large enough volume of material, an incident energetic photon, energy $\hbar\omega$, can produce, in the end, $\hbar\omega/\varepsilon$ hole-electron pairs where ε is the average energy of hole–electron pair production. Experimentally, one finds that ε is remarkably independent of the energy of the incident photon. The number of electrons produced can thus be an accurate guide to the energy of the incident photon and so can be used for the analysis of X-rays and even γ-rays.

All the absorption processes described above have in principle corresponding emission processes, though whether the photons have a reasonable probability of emission depends upon the coupling between the electromagnetic radiation and the electrons as compared to other competing processes. For example, the interaction between conduction electrons and lattice vibrations gives rise to phonon emission rather than photon emission. This rapidly brings the free carriers into thermal equilibrium and reduces the probability of photon emission to negligible proportions (process (a) in Fig. 9.2). With this warning in mind, one may say that the emission of photons with energies greater than the band-gap energy are usually caused by free electrons falling in energy to recombine with free holes in the valence band. On the other hand, exciton states and impurity energy levels give rise to transitions in the electron's energy which can create photons of energy less than the band-gap value. In a semiconductor one expects a broader emission spectrum than that from a single 'line' in an atom where the

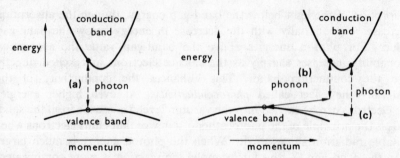

Fig. 9.3 Electron transitions, showing conservation of energy and momentum. (*a*) Direct gap: absorption or emission. (*b*) Indirect gap: photon and phonon emitted on electron–hole recombination, absorbed on electron–hole generation. (*c*) Indirect gap: photon emitted, phonon absorbed on electron–hole recombination; phonon emitted, photon absorbed on electron–hole generation.

electrons have well-separated energy states at clearly defined values. This is because the charge carriers in a semiconductor occupy a range of energies, typically kT wide, in the various bands contributing to emission or absorption. In certain materials, theory adequately predicts the characteristics of absorption or emission, but it can be misleading to draw a generalized absorption or emission characteristic.

Not all electronic transitions from one energy level to another will absorb or emit radiation, even in principle. Some transitions just do not allow photon emission. The permitted transitions are governed by *selection* rules which are more appropriately detailed in a course on quantum mechanics, but two selection rules are worth mentioning. The first, conservation of energy, has already been assumed: an electron changes its energy by the same amount as the emitted or absorbed photon's energy, with usually only a single quantum being concerned in any one interaction. Conservation of momentum in a transition is the second selection rule. The very long wavelengths of photons, as compared to electrons of similar energy, implies that photons have a negligible momentum compared to the electrons. Thus electrons hardly change their k-vector for the emission or absorption of a photon. It follows that in a direct-band-gap material, such as gallium arsenide, photons can be emitted by electrons falling directly from the conduction band into vacant states in the valence band (Fig. 9.3*a*). However, for an indirect band gap material like silicon, such a transition cannot be permitted unless some third particle helps to conserve momentum. In general, the phonon has relatively small energies compared to the band gap, but has values of momentum which cover the entire range for electronic transitions with the crystal. Consequently, in an indirect-band-gap material, photon emission or absorption is accompanied by phonon emission or absorption [Fig. 9.3*b,c*]. One finds that this transition, involving three particles, is less likely than the direct transition which involves only two particles. Thus in general photon emission is more efficient from direct-band-gap materials than from indirect-gap-materials, though, as will be seen later, indirect materials are by no means excluded from forming useful optical devices.

Not all electronic transitions will radiate photons when the electron falls in energy. The emission of phonons has already been mentioned as one method in which an electron can lose energy without radiation being emitted. Another classic process is called Auger recombination and has several forms. The basic effect is that an electron recombines with a hole but the re-combination energy from the transition is given to another electron, possibly freeing it from a bound state into a free state, with the appropriate kinetic energy.

9.3 Photocathodes[9.4,9.5]

When light falls onto a metal or semiconductor surface it is possible for electrons to be emitted from the surface. This is the photoelectric effect and surfaces which show the effect strongly are called *photocathodes*. Such photocathodes are of particular use in some television cameras and image intensifiers. In the image intensifier or image converter (Fig. 9.4) an optical image is focused on to a thin film of photocathode material; electrons are emitted in proportion to the numbers of photons falling on the surface. In this case the film is sufficiently thin for the electrons to be pulled by an electric field into the vacuum behind the film. Electric fields then focus and accelerate the electrons on to a phosphor screen where these energetic electrons liberate many more photons than originally fell on the photo-cathode. The optical image is thus intensified or possibly converted from an infra-red image into a visible range. Frequently there are additional stages. In Fig. 9.4 is shown, schematically, a multi-stage tube where the

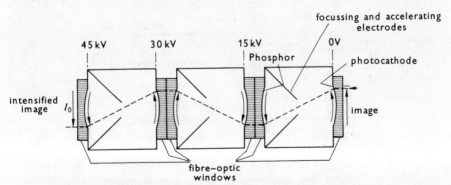

Fig. 9.4 Image intensifier/converter. Picture shows diagram of Mullard XX1060 coupled assembly of image intensifier where light from one phosphor excites the photocathode or the next stage. Coupling of stages is done by fibre optic assemblies. (By courtesy of P. Schagen Mullard Ltd.)

stages are linked by fibre optics. Some other image intensifiers rely on producing electrons from a photo-cathode and then multiplying up the electrons by passing these electrons through arrays of very narrow channels where the electrons produce secondary electrons. Many more electrons then emerge from the channel or tubing than initially entered. Again, at the end of the process the electrons are accelerated into a phosphor screen.

The simplest photoemissive surface is that for a metal. Provided that the incident photon has an energy exceeding the work function of the metal then electrons can gain enough energy to escape from the conduction band into the vacuum level. For photons above the threshold energy, the electron emission is proportional to the light intensity (i.e. proportional to the number of incident photons). The work functions for metals can be low enough to be useful in the optical range (e.g. Caesium has $\phi \sim 1\cdot4$ eV) but the quantum yield (numbers of electrons liberated/number of photons incident) is usually low for the normal optical range. In the first place, the high concentration of free electrons reflects most of the incident light before it can be absorbed by the conduction electrons. Even when an electron has absorbed the photon's energy it will, like all hot electrons in metals, rapidly lose its energy as a result of collisions with the other free electrons and collisions with the crystal lattice. The mean free path for these energetic electrons is of the order of a nanometer in most metals, so that the electron can only escape from the surface if it absorbs the energy of a suitable photon very close to the surface. This distance is too short to give good absorption and thus give a high quantum yield.

The situation can be more favourable in a suitable semiconductor. Figure 9.5a summarizes the points for the photoemission from a bulk semiconductor. The most copious supply of electrons is from the valence band.

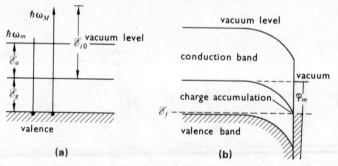

Fig. 9.5 Photoemission from semiconductor. (a) $\mathscr{E}_{i0}$ gives the measure of the threshold energy for impact ionization; $\hbar\omega_m$ is minimum photon energy for photoemission; $\hbar\omega_M$ is maximum energy before secondary ionization by impact starts. (b) Band bending caused by positive surface charge from metal at surface. For lowering of effective electron affinity, band bending must be accomplished within escape depth.

Consequently the most effective photoemission will occur for a photon energy $\hbar\omega > \mathscr{E}_a + \mathscr{E}_g$, the electron affinity plus the band gap energy. However, if the photon's energy is too high then the liberated valence electron has enough energy to ionize an atom and produce a hole–electron pair. This would then cause the liberated electron to lose the energy that it requires for escape from the metal into the vacuum. In order to avoid this possibility one needs

$$\mathscr{E}_g + \mathscr{E}_{io} > \hbar\omega > \mathscr{E}_a + \mathscr{E}_g \qquad\qquad 9.6$$

where $\mathscr{E}_{io}$ is the threshold energy for hole electron pair production. The electrons can of course still lose energy to the free carriers, but this loss process is less severe than in metals which have a much higher density of free electrons. The principal loss of energy to the lattice is through optical phonons, but these typically have energies around only 1/10 eV. Thus in a semiconductor, electrons can escape from as deep as 20 nm below the surface. For photons with energies close to the band-gap energy this distance is still small compared to the characteristic distance of absorption of the photon by the material. Emission is then proportional to this absorption coefficient as well as to the intensity. However, at higher photon energies the absorption length can be small compared to the escape depth, and then the quantum yield is often substantially constant with frequency.

One can see from eqn 9.6 that the material should have a low electron affinity. Semiconductors made from the alkali metals, such as Cs_3Sb and Na_2KSb-Cs, are very effective photocathode materials with quantum efficiencies in the visible range around the 40 per cent region. For example, Cs_3Sb is a semiconductor with $\mathscr{E}_g \sim 1\cdot6$ eV, $\mathscr{E}_a \sim 0\cdot45$ eV and $\mathscr{E}_{io} \sim 1\cdot8$ eV, thus having a threshold photon energy a little over 2 eV. For the lower photon energies around $1\cdot3$ eV (1 micron wavelength) the S1 photocathode, made from Ag-O-Cs, has long been a classic standard, though not well understood in its action. More recently the concept of a zero or negative electron-affinity photocathode[9.6] has come into prominence, and we briefly discuss this concept (Fig. 9.5b).

In section 7.4 it was seen how band bending could be caused by surface charges, arising from surface states, creating a surface potential. In particular a strong downward bending of the bands can be achieved with a p-type material and a strong positive surface charge. To neutralize the surface charge, an accumulation of electrons will form in the conduction band, pinning the Fermi level at the surface close to the conduction band-edge energy. If now, by the use of highly doped material, the change of potential (required to achieve the correct Fermi level in the bulk) is accomplished in a distance that is short compared to the escape depth for the photo-liberated electrons, then the built-in fields will assist the emission of the electrons and so reduce the threshold energy for photoemission. A coating of an electro-positive metal with a low work function is useful for creating the required surface charge and low emission energy. Caesium over gallium arsenside has proved to be very effective, giving quantum efficiencies of over 30 per cent and significant response to wavelengths as long as 0·9 μm, appropriate to the band gap of GaAs. The use of multiple mono-atomic layers of caesium and oxygen has proved even more effective, though less well understood. The use of direct-band-gap material is again considered to be useful in that photons can generate hot electrons in the conduction band without having to have a third particle to take up the charge of the electrons's momentum in the crystal, as is required with an indirect-gap material. For the combi-nation of Cs and GaAs the effective electron affinity is zero, since $\phi_m \sim \mathscr{E}_g \sim 1\cdot4$ eV. Even if the effective electron affinity is negative ($\phi_m < \mathscr{E}_g$), one still often finds that the photoemission is band-gap limited at its lower frequency threshold because the photoliberated electrons first have to enter the conduction band as hot electrons before they can escape from the semi-conductor. The use of this concept of the negative electron affinity appears to have great promise in the engineering of photocathodes that will respond to long wavelengths.

9.4 Photoconductivity[9.2]

Once the photon energy exceeds the band-gap energy then the principal source of absorption for the photons will be through excitation of hole-electron pairs in the valence and conduction bands. The rate of production of holes or electrons is closely linked to the rate at which light is absorbed by the solid. One can define a characteristic length $1/\alpha$ so that the intensity I decays by a fraction α per unit distance of travel of the light through the solid:

$$\frac{dI}{dx} = -\alpha I \qquad 9.7$$

In general α is dependent on frequency and temperature, and sometimes on the intensity itself. If, however, α is independent of the intensity, then $I = I_0 \exp -\alpha x$ where I_0 is the intensity just inside the surface of the semiconductor. The light reflected from the surface of the material is of course lost as far as hole–electron pair production is concerned. Assuming then that every absorbed photon generates a hole and electron (a reasonable approximation for $\hbar\omega \gtrsim \mathscr{E}_g$), one can define a generation rate G_{opt} such that

$$G_{\text{opt}} = \alpha(I/\hbar\omega) = \alpha(I_0/\hbar\omega) \exp -\alpha x \qquad 9.8$$

This gives the rate of generation per unit volume in the material. Note that $\int_0^a G_{\text{opt}} \, dx = I_0/\hbar\omega$, so that, if the material is thick enough to absorb all the photons, there is a corresponding number of electrons (and holes) produced. The absorption coefficients for various materials are shown in Fig. 9.6.

Now, in the dark the density of holes and electrons depends on the impurities in a uniform semiconductor, but in the presence of illumination the density also depends on the generation rate G_{opt} and the recombination rate. The simplest, though not the most realistic, form for the recombina-

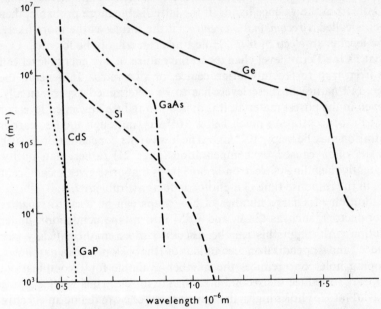

Fig. 9.6 Approximate absorption coefficients for various materials at 300 K.

tion rate in excess of the thermal generation rate is given (page 52) by $R = R_e[(np - n_i^2)/n_i^2]$. If photons also create an additional generation rate G_{opt}, then one can form a rate equation for the conduction band:

$$\frac{dn}{dt} = G_{opt} - R_e[(np - n_i^2)/n_i^2] \qquad 9.9$$

For simplicity, assume a step change in the incident light and charge neutrality ($n = p$) at all times with the initial values of n and p negligible compared to their final values; then eqn 9.9 integrates to give

$$n = n_0 \tanh (G_{opt}t/n_0); \qquad n_0 = n_i[1 + (G_{opt}/R_e)]^{1/2} \qquad 9.10$$

This result is used only to bring out two points. Firstly, a large change in the free-carrier concentration requires a low effective recombination rate

R_e, and secondly the time taken ($\sim n_0/G_{opt}$) to respond to changes in illumination is governed by the time taken to generate the required density of carriers.

Now, even in materials where there are no specific donor or acceptor impurities, one can still find many impurities present that create energy levels somewhere in the band-gap. These can significantly affect the numbers of photo-generated carriers. Electrons which fall into such deep levels can either be thermally stimulated back into the conduction band or recombine with a hole from the valence band. If the latter is the more probable, then the level is called a *recombination centre;* if the former is the more likely, then the level is an *electron trap.* Hole traps can equally be formed. One finds that as the Fermi level changes so the nature of any energy level can change from trap to recombination centre, or vice versa. In general, the presence and nature of these levels has to be determined experimentally. Now, even in the purest material, it is difficult to reduce unwanted levels or centres to a concentration much below $10^{18}/m^3$, although the free-carrier concentration may be below $10^{14}/m^3$. The low carrier concentration in such material is often caused by compensation (page 21) rather than purity. Consequently the filling of electron or hole traps that arise gives a significant increase in the response time of a photoconductor (Problem 9.2).

The addition of a large number of hole traps can be used to sensitize photoconductors, such as GaAs and CdS, into n-type conduction under illumination. Although this can be discussed more mathematically (see Problem 9.2 and Appendix 4 on recombination) the basic physics is as follows. By trapping holes one reduces the number available for recombination. This appears to reduce the recombination rate for electrons while keeping the removal rate for holes high. In such material one can define an effective free-electron time τ_n so that the increase in electron density is $\Delta n = \tau_n G_{opt}$. The increase in current for a block of material, length L and area A, is then

$$\Delta I = e(V/L)\tau_n G_{opt}\mu_n A \qquad 9.11$$

where V is the applied voltage and a uniform illumination is assumed. For a practical device this change of current must be significant compared to the current in the absence of illumination (the *dark current*). This usually requires a high-resistivity material. The ratio of the number of electrons flowing through the external circuit, in unit time, to the number of photons absorbed in the same time defines a quantum gain Q. The electrons flow at a rate $\Delta I/e$ while the photons are absorbed at a rate $G_{opt}LA$, so that, from 9.11:

$$Q = \tau_n/T_n \quad \text{where} \quad T_n = L^2/\mu_n V \qquad 9.12$$

and it can be seen that T_n is the transit time for the electrons through the length L. Quantum gains in excess of unity are possible, though this applies only to material where the contacts are ohmic, so that for every electron leaving at the anode there is one replacing it at the cathode. The effective free time τ_n can then be longer than the transit time because the electron is not lost to the semiconductor once it reaches an electrode. If the electrons leaving at one electrode are not replaced at the other (as for example in a reverse biased p-n junction), then the *effective* free time can never exceed the transit time, no matter how slow the recombination rate. The quantum gain Q cannot then exceed unity.

Applications of photoconductivity can be found in some types of television camera (vidicons), but one finds more and more that these photoconductive effects make better devices when combined with p-n junctions. It is to these photodiodes that we now turn.

9.5 Photodiodes

The photodiode uses the same effects of photoconductivity that were outlined in the last section, but now the generation of hole–electron pairs is encouraged to occur within a depletion region of a p-n junction. In order to obtain close control over the depletion width, or to extend the depletion width, p-i-n diodes are often used. Schottky barrier diodes can also be used. A schematic diagram of a p-i-n diode is shown in Fig. 9.7a. In general, the top contact has to be thin enough to allow the incident light to be transmitted without substantial loss or absorption. One may also have a layer of dielectric material covering the surface of the semiconductor. The purpose of such a layer is to help match the radiation from air or free space into the relatively high refractive index, n, of the semiconductor. Electromagnetic theory shows that one simple form for this 'matching' consists of a layer of relative dielectric constant n^2 and thickness $\lambda/4n$ where λ is the free-space wavelength. This process is comparable to the blooming of lenses in cameras etc so as to cut down the reflected light. Special oxides have been developed to help such matching to semiconductors.

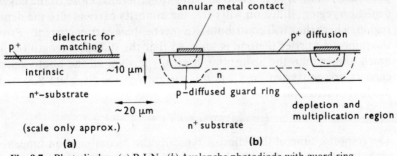

Fig. 9.7   Photodiodes. (*a*) P-I-N. (*b*) Avalanche photodiode with guard ring.

Once the light is in the diode then, as before, eqn 9.8 is assumed to hold for sufficiently energetic photons. If the carrier lifetimes are *short* enough, then the arguments leading to eqn 9.12 can be extended to show that the quantum yield is $[(\tau_n/T_n) + (\tau_p/T_p)]$ where the effect of holes has been added in with an obvious notation. In general, the intrinsic material is chosen so as to have an effective lifetime that is *long* compared to the transit time. There is thus negligible chance of recombination before the optically gener-ated hole or electron reaches its respective contacts. If all the photons are assumed to be absorbed in the depletion region, then one can approach unity quantum gain with the external current given by $e\int G\, dV$, the integral being evaluated throughout the depletion region. The dark current will now be only the reverse saturation current for the junction and can be very small, even for high conductivity material. If the carriers can be removed to the contacts by the field before significant trapping can occur, then the response time is essentially the transit time for the carriers moving across the depletion region. In general, the decrease in dark current and the increase in speed of response more than compensates for the loss of the quantum gain given by the photoconductive effect with ohmic contacts.

If a high quantum gain is required, avalanche multiplication (page 182) has proved to be very effective. Modern technology has enabled silicon diodes to be made with remarkably uniform breakdown characteristics over each area of the junction. Figure 9.7*b* shows such a diode with a special feature of a guard-ring to *prevent* a high edge-field giving edge breakdown in pref-erence to a uniform breakdown over the surface. Multiplication factors as high as 10^4, with uniform quality over the diodes surface, can be made. A single photogenerated electron can then produce thousands more electrons by impact ionization at large reverse-bias fields, and the speed of response is only slightly degraded from the transit time value. Silicon avalanche photodetectors are well matched in wavelength for the detection of the emission from GaAs lasers (discussed in section 9.7).

Photogeneration in the depletion region is not the only mechanism for photodetection. If hole–electron pairs are generated close to the edges of a depletion region, diffusion will drive the minority carriers into the depletion region, where they will contribute to a reverse-bias carrier current. Provided the minority carrier lifetime is long, so that the diffusion length $(D\tau_r)^{1/2}$ is much longer than the absorption length, $1/\alpha$, then all the photo-generated carriers are converted into a photo-current I_{ph} (see Problem 9.3) with the diode characteristic given by

$$I = I_s[\exp(eV/kT) - 1] - I_{ph} \qquad 9.13$$

The response time of this effect is typically the recombination time, so that such detectors are relatively slow; indeed photo-generated minority-carrier storage tends to slow down the speed of response of all p-n junction photo-diodes.

Solar cells

Solar cells are p-n junctions with a long minority-carrier lifetime. The difference in the operation of a photo-diode detector and a solar cell is one of operating point. Figure 9.8a shows this difference of operating point and load line for the photo-diode operated as a detector and operated to utilize the photo-voltaic effect as in a solar cell. The open-circuit voltage developed by a photodiode, with characteristics as in eqn 9.13, is given by

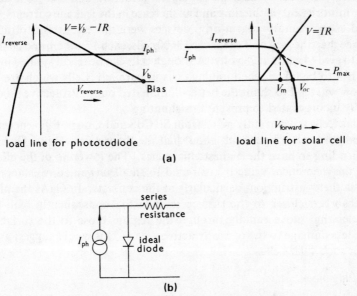

load line for phototodiode   load line for solar cell

(a)

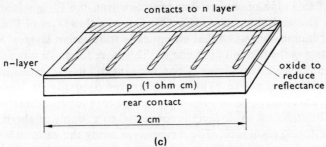

(b)

(c)

Fig. 9.8 A solar cell. (a) Load lines. (b) Equivalent circuit (P_{max} when $V = V_m$; $V_m \sim 0.4$ V for Si). (c) Construction: n-on-p silicon cell.

$V_{oc} \sim (kT/e) \log_e (I_{ph}/I_s)$, while the maximum power is developed at a slightly lower voltage (see Problems 9.4 and Fig. 9.8a). Figure 9.8b shows the equivalent circuit for the solar cell. Here it should be noted that any series resistance arising from construction and contacts must degrade the power and voltage output.

The solar cells for the first commercial satellite Telstar[9.13] were made from p-type Si with a n-type diffused contact (Fig. 9.8c). The reasons for this construction, rather than p on n, raises an important technological point. In the ionosphere, where solar cells cells frequently operate, there are concentrations of charged particles with energies in the MeV range. These particles can actually damage the crystal structure and lead to a decrease in the minority carrier lifetime and an increase in the leakage currents. It was found experimentally that n on p devices were more resistant to radiation damage than the p on n. Telstar used 3600 cells, each 1 × 2 cm and connected so as to give 14 watts at 28 volts. Although cells can be readily joined in series, care has to be taken when joining cells in parallel. Cells which are in any shadow will tend to shunt the better-illuminated ones. Protective diodes are usually incorporated to prevent this shunting.

Solar Cells are currently made from Si, CdS and GaAs. Efficiencies range typically from 8–15 per cent, about half the theoretical value—GaAs solar cells tending to have the highest efficiencies. The lowering of the efficiency from the theoretical value is attributed to small unwanted resistances that arise in the construction, particularly at the contacts. In GaAs the photons are absorbed closer to the surface than in Si; consequently high-energy particles that move rapidly through the regions close to the surface may cause less damage to the optically active areas. This would suggest a longer life for such solar cells.

The Plumbicon[9.7]

A p-i-n diode of large area is used in a very successful television camera developed as an extension of the vidicon. Figure 9.9 shows a schematic section of this type of camera first reported from the Philips laboratories in 1963. The optically active region consists of a thin layer of PbO (hence the name Plumbicon) which has an optically transparent layer of SnO_2 on one side facing the focused light image. This layer forms an n-type contact. The other side of the thin layer faces a vacuum and has an excess of oxygen in order to make it p-type. The region between (typically 10–20 μm thick) forms the intrinsic region of PbO ($\mathscr{E}_g \sim 2 \cdot 0$ eV for the orthorhombic crystal form). The stannous oxide layer is connected to a supply of about 50 volts via a monitoring resistance. The p-type layer facing the vacuum is scanned by an electron beam which charges this p-type face until it drops in potential to the cathode potential, when no further electrons can reach it. Thus the electron beam rapidly reduces the surface of the PbO that faces the vacuum to zero volts—the cathode potential. Now, as light falls on the p-i-n diode the hole and electron pairs that are created drift apart under the action of the reverse-bias voltage (around 50 volts). The holes then drift to the p-type

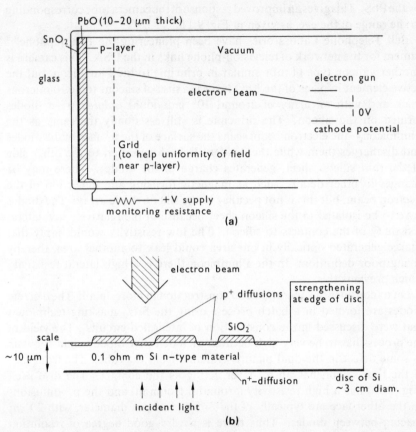

Fig. 9.9 (a) Schematic diagram of Plumbicon. (b) Silicon target construction for silicon vidicon.

surface and charge this surface to a local potential that depends on how much light has fallen on the local area since the beam last scanned or interrogated that area. The charging current is small and does not produce significant voltage across the monitoring resistance, but the rapid discharge of each local area by the beam drives a current (up to a microamp) through the monitoring resistance. The actual current depends on the charge stored in the local area and this in turn depends on the light falling on that area since it was last 'interrogated' by the beam. Television CRTs then have their phosphor-coated screens scanned by electron beams moving in synchronism with the beam in the camera. The intensity of the picture is controlled by the signal from the monitoring resistor in the camera. Later improvements to the Plumbicon added a thin layer of PbS ($\mathscr{E}_g \sim 0.4$ eV) on the electron-beam side of the PbO. Red light that is not absorbed by the PbO is then absorbed by the PbS. This gives an improved response of the camera tube corresponding to the range of the eye, as given in Fig. 9.11.

Bell Telephone Laboratories have been pioneering the Picturephone[9.8] camera for the network of television-phone links in the USA. This camera is another vidicon type of tube similar in principle to the Plumbicon, but the active element in place of the PbO layer is a disc of silicon, some 10 microns thick and with an array of around 10^6 individual *isolated* p^+-n diodes formed on the silicon. The principle is still essentially the same as the Plumbicon—the electron beam scans the surface of the p^+ face of the diodes and discharges them while the incident light, focused on to the other side of the thin silicon sheet, generates charge in the diodes. There may be changes in other details, such as magnetic focusing and deflection of the electron beam, but this is not peculiar to the silicon diode array. The diodes have to be isolated in the silicon sheet because of the relatively low lateral resistivity of the contacts to silicon. The low resistivity would imply that charge generated optically in one area could leak to another area, thereby giving poor definition. In the Plumbicon there is a high lateral resistivity which prevents this.

Let us consider the operation and construction in more detail. The discrete diodes are formed in a batch process using the SiO_2 masking techniques that were discussed in the construction of integrated circuits. The yield of the process has to be very high because each defective diode would show up as some defect in the final picture available from the tube. The front face of the thin silicon has an n^+-layer acting as the contact. The disc itself (Fig. 9.9b) has a high resistivity (around 0.1 ohm m) and the p^+-diffusions on the other face are typically of the order of 10 μm diameter, with 20 μm spacing between diodes. Thus there is a very good degree of resolution available. The depletion region under the p^+-diffusions does not necessarily extend right across to the front face; provided the diffusion length $(D\tau_r)^{1/2}$ for the minority carriers is long enough, diffusion can easily drive the photo-generated carriers into the depletion region, as in the solar cell. The electron beam inevitably scans the oxide surface as well as the p^+-areas and it is found that the oxide tends to charge and produce electric fields which defocus the beam or give other unwanted effects. One technique that is used to reduce this effect is to cover the surface that is scanned, by the beam with a *resistive-sea*. A thin film of resistive material can have a low enough resistivity to reduce the charging effects but still have a high enough resistivity to prevent charge from one diode leaking to another in the time of one scan period.

The dark currents for this type of tube are measured in nanoamps while the sensitivity is remarkably uniform over a range of 0·45 to 0·85 μm wavelength. Values of approximately a microamp per mW/m^2 of incident illumination are obtained, giving final figures around 500 mA/W for the typical areas that are used (in round numbers this gives 1 mA/lumen). The tube is more sensitive than other vidicons, it has a higher tolerance to very high light levels without the photosensitive surface burning out, and the uniformity of sensitivity makes it valuable in color television and other applications.

Particle detectors

As mentioned in section 9.2, X-ray or γ-ray photons can be completely absorbed in large enough volumes of semiconductor. It has proved possible to make p-i-n diodes of silicon (and germanium) in which the i-layer is tens of mm thick and with similarly large lateral dimensions. Such a diode can absorb the energy that is liberated by an X-ray photon and produce close to $(\hbar\omega/\varepsilon)$ hole–electron pairs in the depletion region. For Si, the average energy of hole–electron pair production is given by $\varepsilon = 3\cdot6$ eV ($2\cdot7$ eV for Ge). Germanium diodes give better absorption than Si for γ-rays on account of the heavier mass of the Ge atom. The absorption process is statistical in nature but one finds a smaller random fluctuation in the number of hole–electron pairs per photon than one might expect. If the mean number per photon is n_0 ($= \hbar\omega/\varepsilon$), then the variance is $\overline{(n - n_0)^2} = Fn_0$. For a Poisson distribution F should be unity, but experimentally it is found to be as low as $(1/10)$ or so. F is called the Fano factor, and theoretically is accounted for by the fact that nearly *all* the energy is turned into hole–electron pairs. One can thus make detectors that resolve the energy of energetic photons. Electronic instrumentation has been developed that analyses each small pulse of charge and counts the number of pulses with different charge levels that arise per unit time. Provided, then, that the X-ray photons arrive singly, one can obtain an analysis of the energy of the incident radiation. This can be used in X-ray analysis of materials.

One solution to the problem of producing large volumes of intrinsic Si is to use the fact that Lithium is a donor impurity that can readily move interstitially through Si or Ge at about 100 °C. If Li is diffused into p-type Si it forms a p-n junction. If a reverse bias is applied, the reverse-bias field forces the positively ionized Li atoms to drift into the p-region.[9.14] These ions, acting like charge carriers, will redistribute themselves so as to neutralize the conductor (see section 3.3). Thus the mobile Li ions neutralize the p-type acceptors in the Si, these normal acceptors being immobile at temperatures as low as 100 °C. This produces almost ideal compensated intrinsic material with donors and acceptors in equal quantities. Unfortunately, for ever after—whether in use or not, one must keep the Li-drifted material cooled to liquid-nitrogen temperatures so as to prevent further redistribution of the Li ions. In spite of this, the excellence of these detectors makes them much in demand for X-ray analysis.

9.6 Electroluminescent diodes[9.9 – 9.11]

Electroluminescence is one of the many methods of producing spontaneous emission of photons from a solid. In electroluminescence the numbers of electrons at the higher energies are enhanced above the thermal equilibrium value by forcing a current through the solid. The precise mechanism for populating the higher energy levels in preference to the lower can vary from tunnelling of electrons and avalanche breakdown to injection of charge carriers across a p-n junction. This last possibility is the one considered here. Such electroluminescent p-n diodes can be integrated into arrays with logic circuits which activate the required patterns of diodes to form letters or numbers (alpha-numeric displays). These light-emitting diode (LED) displays find a ready use in many equipments.

The mechanism of electroluminescence in p-n diodes is indicated qualitatively in Fig. 9.10. The p-n junction is made from heavily doped (p^+-n^+) material. On forward bias, electrons become injected into the p-region and holes into the n-region, so that these minority carriers can readily recombine. This spontaneous recombination can be accompanied by the emission of a photon, provided energy and momentum are conserved. As pointed out earlier, there is virtually no change of momentum of the electron in any interaction with the photon because the photon has such a long wavelength (short k-vector). Thus radiative recombination occurs most readily when the conduction band minimum is directly above the valence band maximum in k-space. This is a direct-band-gap material; GaAs is a good example of

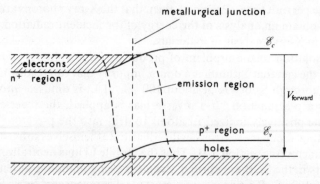

Fig. 9.10 Band diagram (schematic) for forward-biased electroluminescent diode, showing quasi-fermi levels. - - -

such material with infra-red electroluminescence. The energy of the photons emitted corresponds closely to the band-gap energy, though there is a typical spread of approximately kT in this energy; this is because there is a range of energies for the electrons above the conduction-band edge and for holes below the valence-band edge. Thus lowering the temperature reduces the spread of emission. In general, the band gap also changes with temperature, so the central emission wavelength may vary. For example, in GaAs the peak emission is around 900 nm at room temperature but at 840 nm at the temperature of liquid nitrogen. Both these are in the infra-red. For emission in the visible range one requires a higher band-gap material.

One interesting technology that is expanding is the use of ternary semi-conductors. For example, GaP has a band gap of 2·4 eV while GaAs has a band gap of 1·4 eV. The mixture $GaAs_{1-x}P_x$ also forms a semiconductor with a band gap that varies depending on the value of x. Successful red emission diodes have been made and marketed using value of x up to 0·36 approximately. This value cannot go higher successfully because GaP is an indirect-band-gap material. It does have a minimum in its conduction band at about 3 eV directly above its valence band, but this is not the lowest minimum, which is situated 2·4 eV above the valence band at the edge of the Brillouin zone in k-space. The value of x changes the gap of this indirect valley, and at approximately $x = 0·45$ the mixture changes from the direct gap being the lowest energy valley to the indirect gap being the lowest. Thus for values of x which approach this value one finds that the electrons tend to populate the indirect valley in the conduction band and the probability of 'band-gap' radiative recombinations decreases very rapidly. Recombination in indirect material occurs with the aid of phonons rather than photons (Fig. 9.3).

An important quantity is the quantum efficiency of an electroluminescent diode. This is given by the number of photons produced per unit time divided by the number of electrons injected in the same unit of time. There is of course a distinction between the internal efficiency and the external efficiency. The number of photons escaping from the semiconductor is not the same as the number generated internally. As for photodetectors, the difference in the index of refraction for the semiconductor and the atmos-phere outside creates considerable reflection. In fact the situation is worse because of the phenomenon of *total internal reflection*, as used in prisms. While the use of special coatings can help, the shaping of the diode into a hemisphere so that the light falls normal to the interface cuts down on the total internal reflection and greatly enhances the external quantum efficiency. Reabsorption of the generated photons can also occur, *within* the semi-conductor. An effective method of combating this reabsorption is made with $GaAs_{1-x}P_x$ diodes, by changing the value of x away from the junction so as to increase the band gap. The band-gap emission at the junction cannot then be reabsorbed by band-gap absorption away from the junction. However the efficiencies of $GaAs_{1-x}P_x$ diodes are not nearly so good as those of GaAs where recorded external efficiencies at low temperatures have approached 50 per cent.

From what has been said about indirect-band-gap materials one may suppose that they cannot emit light efficiently. This is not the true picture. Efficiencies of several per cent have been obtained with red emission from GaP p-n junctions in spite of the material having an indirect gap. However, here the addition of impurities is of prime importance. One such emission process (around 698 nm) is created by oxygen and zinc impurities forming together a set of localized electronic energies—a *complex*. Such a complex can trap injected electrons at high energies and then allow them to fall to a lower energy state of the complex leading to spontaneous photon emission in the red band. Useful green emission [around 565 nm] has also been made to occur, but in this case the emission is caused by exciton recombination with the exciton bound to a nitrogen impurity. Both processes are rather specialized and so are left for further reading. The green emission is much less efficient than the red in GaP diodes. However, the eye is more sensitive to the green light than to the red. Figure 9.11 indicates a typical

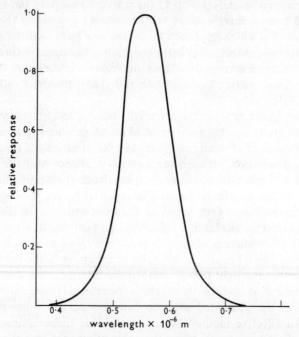

Fig. 9.11   Relative response of eye (good light conditions).

curve for the sensitivity of the eye in good light conditions; the peak sensitivity occurring at 555 nm wavelength. One can therefore tolerate a lower efficiency in the green for the same effective intensity as the deep red.

9.7 The injection laser[9.12 – 9.14]

In the first section of this chapter it was shown that there were two methods of producing photons: spontaneous emission and stimulated emission. The two types of emission have markedly different characteristics in their quality. Stimulated emission is said to be coherent. This means that all the photons that are emitted are in phase with the stimulating radiation. Spontaneous emission is said to be incoherent because the photons have a random phase. The property of coherence is never absolute. All electromagnetic radiation can be described by a propagating electric field such as $E = E(t_0) \cos [\omega t - kz + \phi(t_0)]$ where t_0 is the time of origin. For coherent emission at a frequency ω the field $E(t_0)$ and phase $\phi(t_0)$ are independent of the time t_0. However, all practical radiation has a spread on these parameters but stimulated emission has the amplitude and phase very well stabilized about mean values, E_0 and ϕ_0 respectively. Spontaneous emission has a very much greater spread with time. For any communication purpose where the information is sent in the form of a variation of amplitude or phase with the time of origin t_0, incoherent changes in E_0 and ϕ_0 could be mistaken for signal, or may even mask the signal completely. The discovery† of how to obtain stimulated emission at high frequencies by the correct application of eqn 9.3 thus has considerable importance for the future of optical communications. This discovery applied to optical frequencies has been termed the LASER, an acronym for Light Amplification by Stimulated Emission of Radiation. There are many possible systems that will give laser action but here we shall only consider the injection laser. This uses p-n junctions of materials which show ready spontaneous emission. Successful lasers of this type are at present confined to direct-band-gap materials where there is a high probability of radiative recombination across the band gap.

If spontaneous emission is ignored, then eqn 9.3 can be rearranged to give

$$\frac{dP}{dt} = hfBP(f)(F_2 - F_1) \qquad 9.14$$

The notation of section 9.1 has been retained. It can be seen that to obtain a growth of the power density $P(f)$ one requires the probability of the higher energy being occupied to be greater than the probability of the lower energy being occupied. This requirement on the occupation probability is referred to as *population inversion*. In many laser systems the transitions

† A. L. Schawlow and C. H. Townes *Phys. Rev.* **112**, 1940 (1958).

made by the electrons are between only two or three energy levels. The population inversion need then be specific to only these levels. In the injection laser one finds that the probability of electrons occupying states in the conduction band must exceed the probability of their occupying those in the valence band. Population inversion then occurs over a range of energies close to the band edges. This is achieved, as the following argument will show, by the application of a forward bias to a p-n junction made from highly doped p and n material.

When both sides of a p-n junction are so highly doped that Boltzmann statistics cannot be applied to either side, the technique of quasi-Fermi levels has to be used (section 4.2). In considering non-degenerate p-n junctions, the concentration of, say, the holes on the n-side is given as $N_v \exp -(\mathscr{E}_f - \mathscr{E}_v - eV)/kT$. The Boltzmann distribution of energies is retained but the effective fermi level is lowered, for holes, from $\mathscr{E}_f$ to $\mathscr{E}_{fp} = \mathscr{E}_f - eV$. This assumes no recombination. A similar result holds for the electrons, so that throughout the p-n junction, the quasi-Fermi levels are separated as $\mathscr{E}_{fn} - \mathscr{E}_{fp} = eV$. Similar arguments can be applied to the Fermi-Dirac distribution: ignore recombination in the depletion region and assume a negligible rearrangement of the distribution of energies. The electron Fermi level is then determined from the n-side while the hole fermi level is determined from the p-side, and the two levels are separated by eV, except as recombination restores equilibrium. Figure 9.10 indicates schematically the quasi-Fermi levels for a degenerate p-n junction when these assumptions are made. Returning now to eqn 9.14, it can be seen that to obtain population inversion

$$[1 + \exp(\mathscr{E}_c - \mathscr{E}_{fn})/kT]^{-1} > [1 + \exp(\mathscr{E}_v - \mathscr{E}_{fp})/kT]^{-1}$$

or $$\mathscr{E}_c - \mathscr{E}_v < \mathscr{E}_{fn} - \mathscr{E}_{fp} = eV \qquad 9.15$$

The applied voltage must therefore be greater than the band gap if inversion is to be obtained. If a depletion region is to be maintained, then the material must clearly be strongly degenerate, with the Fermi levels well inside the respective band edges as in Fig. 9.10.

We now turn to considering a typical geometry for a junction laser (Fig. 9.12). Two optically flat and parallel faces are formed perpendicular to the plane of the junction; cleaving the crystal in GaAs usually exposes faces of sufficient flatness and parallelism. The adjacent faces are left rough. To keep the total heat dissipation down to acceptable levels the dimensions of the laser are usually kept extremely small, and etching to thin dimensions is usually required; etching also leaves a suitable rough surface finish for the side faces. Then only the two flat parallel faces form an optical resonator by reflecting the light back and forth. Such reflecting surfaces form a *Fabry-Perot Etalon*. The action of such a resonator can be understood by assuming that each face reflects a fraction R of the incident amplitude of the electric wave and transmits a fraction T. A source, of amplitude E_s, (Fig. 9.11b) gives a forward wave, amplitude E_f, which is reflected back and reaches the source again after reflection from both faces. Allowing for the reflection loss and the phase change along the path of travel, the forward wave returns with an amplitude $E_{f1} = E_f R^2 \exp -j2\pi(2Ln/\lambda)$ where n is the refractive index and λ is the free-space wavelength (λ/n is the wavelength in the medium).

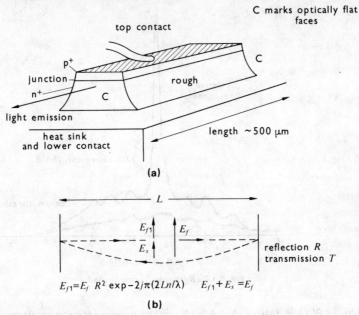

Fig. 9.12 (*a*) Construction of GaAs laser (schematic). (*b*) Action of reflecting faces (Fabry-Perot resonator).

To match the amplitudes, $E_s = E_f\{1 - R^2 \exp[-j2\pi(2Ln/\lambda)]\}$. The forward wave has a maximum amplitude simultaneously with the amplitude, TE_f, of the escaping radiation when

$$L = \tfrac{1}{2}m(\lambda/n): \quad m \text{ integer} \qquad\qquad 9.16$$

Light then is emitted, mainly from the flat end faces, at frequencies where the wavelength satisfies eqn 9.16.

When the current is first passed through this type of device, spontaneous emission will be observed over the usual relatively broad band of frequencies. As the current is increased the observed emission will peak at wavelengths satisfying eqn 9.16, because the internal fields become enhanced by the resonance. Finally, at a threshold current one peak grows to dominant proportions and the device is said to be oscillating by stimulated emission (the term *lasing* is sometimes used). Figure 9.13 shows a characteristic change of output as the current is increased.

The eqn 9.5 can be used to obtain an estimate of the stimulated emission process from the observed spontaneous emission. It is supposed that the spontaneous emission is confined to a well-defined region with width d around the junction, and that the emission has a power density $P(f)$ when a current of density J flows. The emission rate of photons per unit volume is

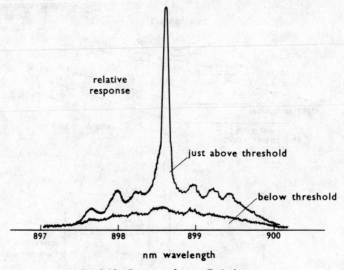

Fig. 9.13   Response from a GaAs laser.

then $\int [P(f)/hf]\, df = N_{sp}\, \Delta f$; N_{sp} is the *average* spontaneous emission rate per unit frequency interval and Δf is the *effective* width of the spontaneous emission. For GaAs $h\, \Delta f \sim 15$ meV at liquid-nitrogen temperatures and ~ 35 meV at room temperature. The value of $N_{sp}\, \Delta f$ can also be related to the current density through the internal quantum efficiency η since the rate of photon emission must be $(\eta J/ed)$ per unit volume $(= N_{sp}\, \Delta f)$. Hence the effective value of $AF_2(1 - F_2)$ is given by N_{sp} the spontaneous emission per unit frequency interval and this means that

$$N_{sp} = AF_2(1 - F_1) = (\eta J/ed\, \Delta f) \qquad\qquad 9.17$$

But from eqn 9.2, and Problem 9.1 the equilibrium stimulated emission density must be given by

$$N_{st} = P(f)(c^2 c_g/h8\pi f^3)(\eta J/ed\, \Delta f) \qquad\qquad 9.18$$

Assuming that the population inversion is complete, so that the stimulated emission dominates, then the greatest rate of increase of $P(f)$ is (using eqn 9.3)

$$\frac{dP}{dt} = \frac{dP}{dt} \cdot \frac{dx}{dt} = c_g \frac{dP}{dx} = c_g(gP); \qquad g = (\eta J c^2 / 8\pi f^2 ed \, \Delta f) \qquad 9.19$$

Thus as the power density travels with the group velocity c_g through the medium it should grow spatially at a rate $\exp gx$. However, this makes no allowance of any distributed loss, α/unit length, and discrete losses at the end faces. Two factors contribute to the distributed loss. Not all the power in the light travels in the single direction of propagation defined by the reflecting faces of the Fabry-Perot resonator. Some power escapes sideways into space or into the semiconductor; this is caused by diffraction, because the wave cannot be a true plane wave over such a small area. A second contribution arises as a result of absorption of light by free carriers, or indeed by reabsorption across the band gap at the regions where the stimulated emission process is weak. Typical overall values to the parameter α are of order 10^3 m^{-1}. The net result is a growth $\exp(g - \alpha)L$ through the length L. To reach the oscillation condition, the overall gain must exceed the overall loss, so that if R^2 is the power loss from each of the reflecting planes it follows that the threshold condition for oscillation is

$$(R^2 \exp(g - \alpha)L) = 1 \quad \text{or} \quad g = \alpha + L^{-1} \log_e R^{-2} \qquad 9.20$$

For oscillation at the lowest current values one requires a high reflection (R close to unity). In principle the device can also be used as an amplifier. Any light of the correct frequency that is directed into the length of the junction will emerge with a larger amplitude provided $\exp(g - \alpha)L$ is large enough. After all the a in *laser* does stand for *amplification*. However, for amplification one requires a low value of R to prevent oscillation.

The result of eqn 9.19 with eqn 9.20 is mainly of use to indicate the important factors for a low starting current for oscillation. Although it can be used to estimate the starting current in some ideal low-temperature cases, it is too inexact (for example the assumption of complete inversion) for general answers. It does show the necessity for a high current density and the desirability of confining the emission to a narrow region (d small). Consideration of the distributed loss suggests that it would be advantageous to help to guide the waves along the junction, and so reduce diffraction losses and absorption losses at the edges of the narrow junction.

The first successful injection lasers were made from GaAs and had threshold current densities of order 4×10^8 A/m^2 for operation at room temperature, but could only operate on duty cycles around 0·1 per cent in pulsed operation. The heat generated could not be extracted without an excessive rise of temperature. Recently, significant advances have been made using ternary semiconductors. $Ga_xAl_{1-x}As$ is a composite semiconductor material with a band gap that varies from 1·4 (pure GaAs) to 2·2 eV (pure AlAs). An important new feature made available by modern epitaxial technology is that one can grow this material with various values of x directly onto GaAs. This growth forms junctions between two materials with different band gaps—*heterojunctions*. Junctions of this form have been known about theoretically for a long time. Heterojunctions between p and n type materials can give high injection efficiencies to one type of carrier by making the opposite charge carrier cross a very high potential barrier. This property of the extra potential barrier created by a dissimilar band gap is used to good effect in the heterojunction laser (Fig. 9.14). The concept of quasi-Fermi levels has been used to indicate the level of holes and electrons in the junction on forward bias. One can see clearly how the change to a wider band-gap material on either side of the GaAs p-layer provides a potential barrier that tends to confine the injected charge carriers to the central region. The $Ga_xAl_{1-x}As$ thus provides the higher band-gap material which confines the charge carriers and hence the emission to a width defined by the central

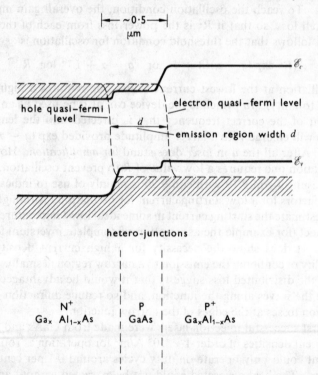

Fig. 9.14 Simplified band diagram of GaAlAs heterojunction laser (on forward bias). End contacts of p⁺ and n⁺ GaAs are at present used after the GaAlAs sections; these allow better contact to the metal electrodes.

GaAs region. This can be made in the sub-micrometre region, thus helping to lower the threshold current substantially. Additional benefits are obtained from the double hetero-junction structure in that the refractive index for GaAs and $Ga_xAl_{1-x}As$ are slightly different. This helps to guide the electromagnetic fields along the junction and reduce the diffraction loss. The higher band gap on either side of the junction also helps to reduce any absorption across the band gap at the edges of the emission region. With these techniques it has proved possible to operate the injection laser continuously, the threshold currents now falling to around 10^7 A/m^2.

For communication purposes one needs to consider how the device could be modulated. Changing the current once the device is producing stimulated emission provides a response that is well into the microwave frequency range. However, switching the current from zero into values above threshold takes several nanoseconds to produce the light. The reason for this much slower response to switching-on comes from the reliance on spontaneous emission to start the oscillation. The time constants for spontaneous emission are typically the recombination time constants for the charge carriers, and these are known to be in the nanosecond range for GaAs.

The room-temperature semiconductor laser is thus an appropriate topic on which to end this book. It is a device which utilizes to the full the range of technologies that the semiconductor manufacturer has at his disposal and it has promise for the future. Indeed, it is envisaged that modulated laser light, propagated along microscopic fibres of glass, will be able to carry hundreds of television channels, or thousands of telephone links, between major centres of population. The task will not be easy, but to the combined skill of chemists, physicists and engineers it will prove a challenge that will surely have many useful results.

PROBLEMS

9.1 Show that $\dfrac{\text{Probability of spontaneous emission}}{\text{Probability of stimulated emission}} = \exp \dfrac{hf}{kT} - 1$

Estimate the frequency at which both processes are equally probable at room temperature.

9.2 Read the Appendix on recombination. Consider an ideal material with a direct band-to-band recombination process as on page 52 but with P_t/m^3 ideal hole-traps with zero capture cross-section for electrons from conduction band. Form rate equation $(df/dt) = -vC_p[pf - (1 - f)p_1]$ for occupation fraction f of the energy level of hole trap. Show for charge neutrality $p + (1 - f)P_t = n$. Assume $p_1 \gg p$ and $P_t \gg p_1$. Hence show from rate equation for conduction band that in equilibrium with a photogeneration rate G_{opt} that $n = n_i(G_{opt}/R_e^*)^{1/2}$ where $R_e^* = R_e(p_1/p_t)$. Hence explain how hole traps effectively reduce recombination rate, and sensitize material into n-type conduction.

Now add an additional impurity with density N_t and assume that these are ideal electron traps. Use assumptions similar to those above to show that occupation of the electron traps is approximately $n(N_t/n_1)$ while occupation of the hole traps is $p(P_t/p_1)$; explain what is meant by p_1 and n_1 in this context with the two levels of impurity. Show that for charge neutrality $n + n(N_t/n_1) = p + p(P_t/p_1)$. Hence show that sensitization into an n-type photoconductor is still possible provided $P_t \gg N_t$ and that the effective recombination parameter is $R_e^* = R_e(N_t n_1/P_t p_1)$. Show also that the generation rate has to have an approximate time constant $n_0(N_t/n_1 G_{opt})$ where n_0 is the number of free electrons produced (much smaller than the number of trapped electrons assumed).

9.3 An ideal n-on-p solar cell is assumed to have its junction very close to the surface, so that there is negligible absorption of light until the p-region. At the point x_0 the generation rate is $G_{opt}(x_0)$; give an expression for this in terms of the incident light intensity I_0 at $x = 0$. At the point x_0, G_{opt} contributes a minority carrier density $\Delta n(x_0)$ to the actual minority carrier density $n(x_0)$. This particular contribution then leads to contributions $\delta n(x)$ at other points of x. Show that

$$\delta n(x) = \Delta n(x_0)[\sinh{(x/L_n)}/\sinh{(x_0/L_n)}] \qquad 0 < x < x_0$$
$$= \Delta n(x_0)\exp{(-x/L_n)} \qquad x_0 < x$$

($x = 0$ is taken as the edge of the depletion region where the fields can remove the minority carriers readily). Show how to obtain a balance of generation and removal so

$$(D\,\Delta n(x_0)/L_n)[\cosh{(x_0/L_n)} + \sinh{(x_0/L_n)}] = eG_{opt}\sinh{(x_0/L_n)}$$

Hence by superposition of all the contributions of $\Delta n(x_0)\,dx_0$ from $x_0 = 0$ to $x_0 = \infty$ find current

$$D(\partial n/\partial x)_{x=0} = \int (D/L_n)\frac{n(x_0)}{\sinh{(x_0/L_n)}}\,dx_0$$
$$= e(I_0/h\omega)[\alpha L_n/(1 + \alpha L_n)]$$

Explain the significance of this current, and the significance if $\alpha L_n \gg 1$.

9.4 Find maximum power condition for solar cell $(\partial P/\partial V = 0)$ and show

$$(eV_m/kT)\exp{(eV_m/kT)} \simeq (I_{ph}/I_s)$$

Hence show $V_m \sim V_{oc} - (kT/e)\log_e{(eV_{oc}/kT)}$ for maximum power.

9.5 Photons at the peak sensitivity of the eye (555 nm) create approximately 680 lumens per watt. A red electroluminescent diode carries 20 mA and has a luminous intensity of 10^{-3} candela. Estimate the external quantum efficiency. (1 candela gives out a luminous flux of 4π lumens.)

General references

9.1 PANKOVE, J. I. *Optical Processes in Semiconductors*. Prentice-Hall, 1971.
9.2 LARACH S. (editor). *Photoelectronic Materials and Devices*. Van Nostrand, 1965.
9.3 SZE, S. M. Reference 1.5, Chapters 12 and 13.

Special references

Photocathodes, vidicons etc.

9.4 BECK, A. H. W. and AHMED, H. Reference 1.1, Chapter 7.
9.5 KAZAN, B. and KNOLL, M. *Electronic Image Storage*. Academic Press, 1968.
KAZAN, B. Image Pick-up and Display Devices, in *Topics in Solid State and Quantum Electronics* (editor W. D. Hershberger). Wiley, 1972.
9.6 SCHEER, J. J. and VAN LAAR, J. GaAs-Cs: A new type of photoemitter. *Solid State Communications*, 3 (1965), 189–193.
9.7 HAAN, E. F., VAN DER DRIFT, A. and SCHAMPERS, P. P. M. The Plumbicon. *Phillips Tech. Rev.*, 25 (1964), 130–180.

9.8   CROWELL, M. H. and LADUBA, E. F. The Silicon diode array camera tube. *Bell System Tech. J.*, **48** (1969), 1481–1528.

Light-emitting diodes

9.9   BERGH, A. and DEAN, P. J. Light-emitting diodes. *Proc. IEEE*, **60** (1972), 156–223.
9.10 IVEY, H. F. Electroluminescence and related effects. In *Advances in Electronics and Electron Physics*. Academic Press, 1963.
9.11 THORNTON, P. R. *Physics of Electroluminescent Devices*. Spon, 1967.

Lasers

9.12 SMITH, W. V. and SOROKIN, P. P. *The Laser*. McGraw-Hill, 1966.
9.13 GOOCH, C. H. (editor). *Gallium Arsenide Lasers*. Wiley, 1969.
      GOOCH, C. H. *Injection Electroluminescent Devices*. Wiley, 1973.
9.14 GOODWIN, A. R. and SELWAY, P. R. Heterostructure injection lasers. *Electrical Communication*, **47** (1972), 49–53.

Special items

9.13 SMITH, K. D., GUMMEL, H. K., BODE, J. D., CUTRISS, B. D., NIELSEN, R. J. and ROSENZWEIG, W. Solar cells and their mounting. *Bell System Tech. J.*, **42** (1963), 1765–1816.
      Special issue on Telstar.
9.14 PELL, E. M. Ion drift in n-p junctions. *J. Appl. Phys.*, **31** (1960), 291–302.
      An account of Li drift techniques.
9.15 LOUISELL, W. H. *Radiation and Noise in Quantum Electronics*. McGraw-Hill, 1964.

Appendices

APPENDIX 1 DIFFUSION OF IMPURITIES[2.10,2.11]

Impurities cannot be created or destroyed once inside the material, and so the flux of impurities F across a plane is related to their concentration C by a continuity relationship (as for electrons but without the charge, see eqn 3.8). Thus $\partial F/\partial x = -(\partial C/\partial t)$ and combined with eqn 2.1 (Fick's first law) gives Fick's second law

$$D\frac{\partial^2 C}{\partial x^2} = \frac{\partial C}{\partial t}$$

The powerful method for attacking this type of equation is that of the Laplace transform $\bar{C}(s, x) = \int_0^\infty C(x, t) \exp -st\, dt$. Initially, in this example, there are zero impurities, except at the edge $x = 0$. Thus

$$s\bar{C} = D\, \partial^2\bar{C}/\partial x^2 \quad \text{with} \quad \bar{C} = \bar{C}_1(s) \exp -(s/D)^{1/2}x$$

(i) when $C(t, 0) = C_s$ for all time it follows that $\bar{C}_1(s) = C_s/s$, i.e. a step function of the surface concentration C_s is applied at the surface and maintained at $x = 0$ for all time. This is the first solution for the problem and inversion of the Laplace function† $C = (C_s/s) \exp -(s/D)^{1/2}x$ yields

$$C = C_s \frac{\pi x}{D^{1/2}} \int^t \frac{\exp -(x^2/4Dt)}{2(\pi t)^{3/2}}\, dt$$

$$= C_s \frac{2}{\pi^{1/2}} \int_\phi^\infty \exp -\theta^2\, d\theta \quad \text{where } \phi^2 = x^2/4Dt$$

$$= C_s\, \text{erfc}\, (x/2\sqrt{Dt})$$

(ii) When there is a fixed number of impurities to be redistributed

$$\int_0^\infty C(x, t)\, dx = Q \quad \text{for all time}$$

Hence $\int_0^\infty \bar{C}\,dx = Q/s$ and so $\bar{C}_1(s) = Q/(sD)^{1/2}$. Again, from inversion tables of the Laplace transform†

$$C = \frac{Q}{(\pi Dt)^{1/2}} \exp -(x^2/4Dt)$$

APPENDIX 2   THE ASSESSMENT OF EPITAXIAL MATERIAL[2.9]

A most important aspect in the production of good devices is the control and measurement of the epitaxial layers which form the active devices. We need to know the thickness, the impurity level and the mobility as three major items, though we may also require other information such as the minority-carrier lifetime, density of defects and deep impurities.

The correct growth of an epilayer usually demands a carefully oriented slice of substrate with the surface almost perpendicular to one of the important axes of crystal symmetry. X-ray diffraction patterns are used to align the substrate crystal in a special holder, called a goniometer, before it is sliced into discs and polished ready for the growth of an epilayer. The determination of the thickness of the epilayer is the next problem. With care, the disc of the crystal can be cleaved without too much damage to the top surface, so that preferential etches can be used on the side, cleaved face so as to delineate the boundary between the epilayer and substrate. The thickness of the epilayer can then be measured under a microscope. If the epilayer is very thin then the angle lap technique (page 29) may be required. Other techniques for measuring the thickness make use of the reflection of infra-red light from the substrate and the top of the epilayer—the interference patterns depend on the thickness.

For measuring the conductivity and the mobility, the favoured methods are the *four-point probe* or the *Van der Pauw technique*.[A.1] The former method is straightforward. A current is injected between two probes along a bar of material. The current causes a potential drop to appear along the length. This potential drop is monitored by two more probes (Fig. A.1a). From the geometry and the dimensions of the sample the conductivity may be determined. In all these techniques the epilayer must be grown on a suitable substrate that does not short out the current flow; i.e. a high resistivity substrate or possibly a p-layer on n-substrate, or vice-versa. The measurement of the mobility can also be done with a special sample (Fig. A.1b) which is inserted in a magnetic field and has its Hall voltage

---

† $[C(t) = (\exp -a^2/4t)/(\pi t)^{1/2} \Leftrightarrow \bar{C}(s) = [\exp -a(s)^{1/2}]/(s)^{1/2}$

$\quad C(t) = (\exp -a^2/4t)/2(\pi t)^{3/2} \Leftrightarrow \bar{C}(s) = [\exp -a(s)^{1/2}]/\pi a$

$$\int f(t)\,dt \Leftrightarrow \frac{1}{s}\overline{F(s)}]$$

(Appendix 3) measured across the pairs of electrodes BD and EF when the current is flowing through the electrodes AC.

The Van der Pauw technique combines the measurements of mobility and conductivity in a single specimen (Fig. A.1c) where thickness is the only dimension measured. The technique uses a basic theorem in complex variable theory, and because of its widespread use we outline it here.

Consider first a single electrode injecting a current I into an infinite plane

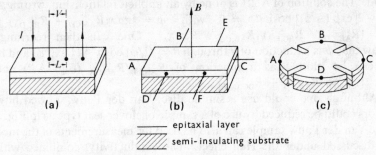

(a)   (b)   (c)

——— epitaxial layer

▨▨ semi-insulating substrate

Fig. A.1 (a) Resistivity measurement. (b) Hall sample. (c) Van der Pauw sample.

of thickness w and conductivity σ. The potential is given by

$$V = (I/2\pi\sigma w) \log r = (I/2\pi\sigma w)\mathscr{R}(\log z)$$

The complex function $\log z$ is referred to as the complex potential. If the electrode is along the edge of a *semi*-infinite plane then the complex potential becomes

$$V = (I/\pi\sigma w) \log z$$

Consider four such electrodes along the straight edge of a semi-infinite plane, and labelled A, B, C, and D with positions arbitrarily taken as $0, b, c, d$, in distance (the reference point being A). A current I is injected into A and taken out of B while the voltage $V_D - V_C$ is measured.

$$V_D - V_C = (I/\pi\sigma w)\{\log [c/(c - b)] - \log [d/(d - b)]\}$$

Define $R_{AB,CD} = (V_D - V_C)/I_{(AB)} = (1/\pi\sigma w) \log [c(d - b)/d(c - b)]$. Similarly if a current is fed in through B and out of C and the voltage is measured across DA then $R_{BC,DA} = (1/\pi\sigma w) \log [c(d - b)/b(d - c)]$. In principle it can be seen that one could do an experiment and measure these resistances $R_{AB,CD}$ etc. and a knowledge of the dimensions would then yield the conductivity. However Van der Pauw was to notice two items.

First, a measurement of two 'resistances' enabled one to eliminate the dimensions because

$$\exp - (\pi\sigma w R_{AB,CD}) + \exp - (\pi\sigma w R_{BC,DA}) =$$
$$[(c - b)d + (d - c)b]/c(d - b) = 1 \quad \text{A.2.1}$$

Secondly, the properties of conformal transformation are well known and a solution for a potential problem in the z plane with a boundary $y = 0$ ($z = x + jy$) can be transformed into a solution for the boundary $w = F^{-1}(x)$ with the potential changing from $\log z$ to $\log [F'(w)]$. The function F can be any function of a complex variable, so that it is possible, in principle, to map the straight edge into any boundary, even the curious shape of Fig. A.1c. The properties of $R_{AB,CD}$ and $R_{BC,DA}$ along with eqn A.2.1 still hold. The solution of A.2.1 is done by an implicit relationship, writing A.2.1 as $2[\exp(-\phi)] \cosh \theta\phi = 1$ with $\phi = \frac{1}{2}\pi\sigma w(R_{AB,CD} + R_{BC,DA})$ and $\theta = (R_{AB,CD} - R_{BC,DA})/(R_{AB,CD} + R_{BC,DA})$. One can then rearrange to show that ϕ is a function of θ through $\theta = (1/\phi) \cosh^{-1}(\frac{1}{2}\exp \phi)$ and hence ϕ can be plotted as a function of $R_{AB,CD}/R_{BC,DA}$ (Fig. A.2) and so $\sigma = [2\phi/\pi w(R_{AB,CD} + R_{BC,DA})]$.

Although one could use a simple disc, Van der Pauw showed how the errors could be reduced by use of a sample of clover-leaf type as in Fig. A.1c. The Van der Pauw sample can also be used for measurements of the mobility as discussed under the Hall effect. The conductivity, combined with the mobility, gives the carrier concentration, at least averaged over the sample.

The Schottky barrier diode has been another source of a measurement technique for the carrier concentration, especially in n-type samples. The capacitance of a depleted region with depletion depth d is $C = A\varepsilon_r\varepsilon_0/d$. Thus a measurement of the capacitance (using r.f. voltage changes that are small compared to the voltage supported by the depletion layer) determines the depth to which the depletion has been driven by a given reverse bias voltage V. If now the steady bias is increased so as to increase the depletion width by δd, then the field is raised everywhere by $\delta E = eN_i(d) \, \delta d/\varepsilon_0\varepsilon_r$; the increase resulting from the additional ionized impurity charge that is uncovered. It can be seen (Fig. A.3) that the field retains its shape (as determined by Gauss' theorem), so that the extra voltage is $\delta V = eN_i(d) \cdot d \cdot \delta d/\varepsilon_0\varepsilon_r$. Eliminating d for C

$$-(dV/dC)(C^3/eA^2\varepsilon_0\varepsilon_r) = N_i(d) \qquad \text{with } d = C/A\varepsilon_0\varepsilon_r$$

Consequently, a careful measurement of the depletion capacitance with voltage in a p^+-n, n^+-p or Schottky barrier diode can yield detailed information about the uniformity of the epitaxial layer as a function of depth. These C-V measurements can be automated with the help of analogue computers or digital capacitance measurements and digital computation. One of the simpler methods for automation has been developed by the author (reference A.2). Such measurements yield useful information on the formation of epilayers and the effects of non-uniformities on device operation. These techniques can be extended to measure impurities that lie deep in the band gap and are only ionized by the action of incident light of suitable

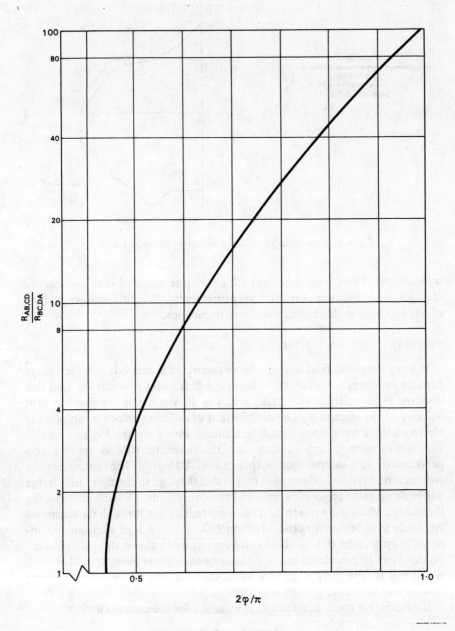

Fig. A.2   Van der Pauw function.

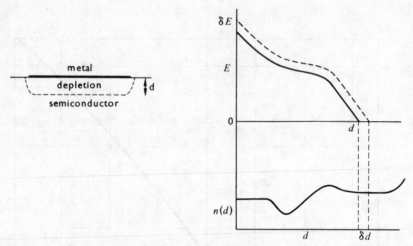

Fig. A.3 Determination of donor density by Schottky barrier.

wavelength. Thus, with a Schottky barrier, one can find that capacitance changes with illumination, and measurements of such photo-capacitive effects can help to determine the 'deep' impurities.

APPENDIX 3 THE HALL EFFECT[3.2,2.1]

When a magnetic field is applied to a sample of semiconductor the charge carriers experience a force $e(\mathbf{v} \times \mathbf{B})$. To a first order one can say that this appears as an equivalent electric field $\bar{\mathbf{v}} \times \mathbf{B}$ where $\bar{\mathbf{v}}$ is the average drift velocity of the electrons. The distribution of different velocities among the electrons does have some significant effects which we shall ignore in this simple treatment. Suppose then that the magnetic field is applied perpendicularly to a sample such as that in Fig. A.1*b* or *c*. The charge carriers will become deflected sideways as they travel through the sample, and charge will build up at the edges of the material and so create an electric field opposing the change which gave rise to it. Thus a current driven through the terminals AC leads to a field across the direction BD. When a high-impedance voltmeter is applied to BD this field can be measured; this is the *Hall* voltage. Its value can be calculated easily if one assumes a plane geometry. However, we intend to give here a slightly more sophisticated version which shows how to find the value for any geometry.

If an effective mean free time τ_m is assumed for the charge carriers

$$\bar{\mathbf{v}} = (e/m)(\mathbf{E} + \bar{\mathbf{v}} \times \mathbf{B})\tau_m$$

For the case where **B** is perpendicular to the current flow (perpendicular to a thin slice of semiconductor) then $\bar{\mathbf{v}} \cdot \mathbf{B} = 0$ and it may be verified that

$$\bar{\mathbf{v}}[1 + (\omega_c \tau_m)^2] = (e/m)\tau_m[\mathbf{E} + \mathbf{E} \times \boldsymbol{\omega}_c] \quad \text{with} \quad (e/m)\mathbf{B} = \boldsymbol{\omega}_c \quad \text{A.4.1}$$

Now when we consider the distribution of the current density $en_0\bar{\mathbf{v}} = \mathbf{J}$ in a material, we have, for the steady state, Div $\mathbf{J} = 0$ and curl $\mathbf{E} = 0$. This latter equation with A.4.1 shows that curl $\bar{\mathbf{v}} = 0$ whether or not there is any magnetic field present. Now it is possible to show that the boundary conditions, combined with curl $\mathbf{J} = 0$ and Div $\mathbf{J} = 0$, uniquely specify the current density distribution in the steady state. Consequently, no matter how complicated the shape of the disc, such as an Van der Pauw sample, provided that the magnetic field is normal to the disc, the distribution of current is *not* changed by the presence of the magnetic field. Since the current flow is not changed, one can use the original current density $J = en_0v$ to find the Hall voltage built up between the two points C and D

$$V_H = \int_C^D (\mathbf{v} \times \mathbf{B}) \cdot dl = \int_C^D (d\mathbf{l} \times \mathbf{v}) \cdot \mathbf{B} = \frac{r\mathbf{B}}{en_0} \cdot \int_C^D d\mathbf{l} \times \mathbf{J} = \frac{rIB}{en_0 D}$$

where I is the current flowing across any line joining the points C and D, and n_0 is the carrier concentration. The factor r would be 1 for the presentation given here, but more detailed calculations (where a spread in value of τ_m with the different carrier velocities among the change carriers is considered) show that $r = \int \overline{\tau_m^2}/(\overline{\tau_m})^2]$. For pure material, dominated by phonon scattering, the value of r is typically close to $1 \cdot 2$; if impurity scattering dominates it tends to 2.

$R_H = r/en_0$ is called the Hall constant and defines the Hall mobility through $\mu_H = R_H\sigma$ where σ is the conductivity.

If one defines $R_{AB.CD} = V_{CD}/I_{AB}$ then on applying $\mathbf{B}$,

$$\Delta R_{AB.CD} = \Delta V_H/I = R_H B$$

A positive value of R_H denotes p-type charge carriers while a negative value denotes n-type charge carriers. This result follows immediately from the directions in which the magnetic field tends to force the charge of a given sign. In near-intrinsic material one may, of course, have significant amounts of both carriers but this is not an example that we consider.

A practical measurement using a Van der Pauw specimen would be made by reversing the magnetic field and noting the change of potential across the appropriate electrodes. The current would then be reversed and the process repeated to ensure an averaging out of any contact potentials.

APPENDIX 4 RECOMBINATION AND TRAPPING

On page 143 the concept of a collision cross-section between two particles was introduced. Here a capture cross-section is introduced. Suppose that an impurity creates an energy state with a level situated in the band gap; one can then assert that C_n is its capture cross-section for an electron from the conduction band. A conduction electron moving with a velocity v will then sweep out an interaction volume vC_n/unit time. All empty levels in question, contained within this volume, will contribute to the capture of electrons, so that if there are N_t impurity levels/unit volume and n electrons/unit volume then the rate of capture will be $nN_t(1 - f)vC_n$ per unit volume. The factor f is the fraction of the N_t levels that are full, so $(1 - f)$ is the fraction that are empty. The velocity v must now be the average thermal velocity. The capture cross-section can also be used in the simple recombination theory (page 51), where one could write $R_e/n_i^2 = vC$, with C as the capture cross-section for conduction electrons by holes in the valence band.

In a similar manner, one can define an escape cross-section D so that the fN_t electrons in the energy levels in the band gap can escape with the average thermal velocity v to the $(N_c - n)$ vacant sites in the conduction band at a rate $vDfN_t(N_c - n)$. It is, however, more convenient to write $DN_c = C_nn_1$ where C_n is the capture cross-section for the energy level and n_1 is defined by (DN_c/C_n). At a later stage n_1 will be given an additional meaning. One notes that $N_c \gg n$. A rate equation may now be established for the conduction band between electrons coming and going from this level as well as direct recombination with the valence band

$$\frac{dn}{dt} = -vC_nN_t[n(1 - f) - fn_1] - (R_e/n_i^2)(np - n_i^2)$$

The valence band similarly exchanges holes with this energy level in question, and has a capture cross section with the level given by C_p. The same formalism holds, and with an obvious notation one may write

$$\frac{dp}{dt} = -vC_pN_t[pf - (1 - f)p_1] - (R_e/n_i^2)(np - n_i^2)$$

Note that if the Fermi level lay at the energy level of the impurity then $f = \frac{1}{2}$. But in equilibrium $dn/dt = dp/dt = 0$ and $np = n_i^2$, so that it follows $n = n_1, p = p_1$ with $n_1p_1 = n_i^2$. This gives a new meaning to n_1 and p_1 as the equilibrium densities of electrons and holes if the Fermi level lies at the energy of the impurity level in question.

A rate equation can also be formed for the electrons in the impurity level:

$$N_t\frac{df}{dt} = vC_nN_t[n(1 - f) - fn_1] - vC_pN_t[pf - (1 - f)p_1]$$

In equilibrium the right-hand side must vanish, and this relationship can be used to evaluate the equilibrium value of f in terms of C_p, C_n, n_1 and p_1.

After some rearrangement and use of $p_1 n_1 = n_i^2$, one can assert for the conduction band:

$$\frac{dn}{dt} = -(np - n_i^2)\left\{\frac{R_e}{n_i^2} + \frac{vC_pC_nN_t}{C_p(p + n_1) + C_n(n + n_1)}\right\}$$

This shows that the addition of an impurity in the band gap increases the rate of recombination from the direct-band-gap value to an effective value $R_e^* \simeq [R_e + N_t vC_p n_i^2/p)]$ if $n \ll p$ and $[R_e + N_t vC_n n_i^2/n]$ if $p \ll n$. Reference back to the minority carrier lifetimes (page 52) shows then that the lifetime for holes in n-type material can be different from electrons in p-type material.

The theory can be extended to show the difference between recombination centres and traps. For the N_t impurity centres to trap electrons the rate of thermal reionization to the conduction band ($vC_n N_t f n_1$) must be greater than the rate of recombination by holes from the valence band ($vC_p N_t f p$). This means that, for an electron trap, $p < C_n n_1/C_p$. Similarly for a hole trap, $n < C_p p_1/C_n$. For a recombination centre the inequalities just expressed must be the other way around. It follows that the energy levels for recombination centres tend to lie in the middle of the band gap, so that n_1 and p_1 are both close to the intrinsic level and are therefore small. Electron traps have the level closer to the conduction band; hole traps are closer to the valence band.

APPENDIX 5   RELAXATION SEMICONDUCTORS[A.3]

In most of the semiconductor theory presented in this book, the dielectric relaxation time τ_d has been assumed to be short, so that electrical neutrality occurs except in depletion regions at junctions, or in regions of negative differential mobility. Thus, when minority carriers are injected into a semiconductor the rapid movement of the majority carriers is assumed to neutralize the excess charge immediately, and the minority carriers then recombine with a distribution in space following the exponential law $\exp -x/l_d$ where $l_d = \sqrt{D\tau_r}$ with τ_r an effective recombination time. The essential approximation here is that $\tau_r \gg \tau_d$. Now there are certain classes of semiconductor where $\tau_r \ll \tau_d$. An example is given by semi-insulating GaAs, where $\tau_r \sim 1$ ns and $\tau_d \sim 100$ μs. Other examples are readily found in the field of *amorphous semiconductors*.[A.4] In these materials, the valence electrons have a reasonably well-defined binding energy to their parent atoms, but there is no long-range order in the periodicity of the atomic sites that allows the clean energy band structure of the perfect crystalline material. The situation is analogous to a coupled set of radio-frequency filters. If all the filters are identical, then the coupling leads to a band of allowed frequencies of propagation with well-defined cut-off points. If the filters are all slightly different, then the permitted pass band of the combined filter is broadened from the ideal; some propagation can take place outside the 'pass-band' and the propagation inside the pass-band is degraded. Thus, in amorphous semiconductors there are energy levels throughout the nominal energy band. The mobility for the material is low because the energy states, at sufficiently close range to the electrons, are not necessarily available. Moreover, the lack of periodicity in the electric potentials in the crystal, a direct consequence of lack of periodicity in atomic spacings, leads to electron waves being readily scattered, another cause of low mobility. The low mobility leads then to a long dielectric relaxation time τ_d. The energy states throughout the nominal band gap also cause rapid recombination, and so it is easy to find that $\tau_r \ll \tau_d$. This inequality characterizes *relaxation semiconductors*. Quite different physics holds for these semiconductors, and it is now shown how injection of minority carriers from a contact can drive the semiconductor into a state of maximal resistivity rather than enhance the conductivity as can happen in the more normal case.

The equations governing the motion are given by 3.13a and 3.13b along with the recombination equations for the steady state, which can be written as

$$\partial J_{cp}/\partial x = -\partial J_{cn}/\partial x = -(np - n_i^2)(\mu_n\mu_p)^{1/2}/(\mu_n n_0 + \mu_p p_0)\tau_r$$

where n_0 and p_0 are equilibrium values and τ_r is an effective recombination time. Considerable clarity is obtained by normalization of the equations governing the motion; the required normalizations are: $(J_{cp} + J_{cn})/J_0 = J$ where $J_0 = em_0kT/eL_d$ with $m_0 = (\mu_p p_0 + \mu_n n_0)$ and $L_d^2 = \varepsilon_r\varepsilon_0 kT(\mu_n\mu_p)^{1/2}/m_0 e^2$

$$M = (\mu_p p + \mu_n n)/m_0; \quad Q = (\mu_p p - \mu_n n)/m_0; \quad Q_c = (\mu_n\mu_p)^{1/2}(p_0 - n_0)/m_0$$
$$F = eL_d/kT \quad X = x/L_d \quad R = \tau_d/2\tau_r \quad \text{where} \quad \tau_d = \varepsilon_r\varepsilon_0/em_0$$

The dimension L_d has the significance of the Debye length that was met in Chapter 3 as the characteristic distance over which charge would readjust itself by diffusion. The time constant τ_d is the dielectric relaxation time for the material in the equilibrium state. In this material R is considered to be very large. Significant numbers of holes and electrons are assumed to be coexisting in this material so that one cannot assume just one type of carrier. The eqns 3.13a and 3.13b can be added together to find the combined conduction J, or they can be used to evaluate $\partial(J_{cp} - J_{cn})/\partial x = -2(np - n_i^2)(\mu_n\mu_p)^{1/2}/\tau_r m_0$. The required equations become:

$$J = MF - \frac{\partial Q}{\partial X}; \qquad \frac{\partial}{\partial X}\left(QF - \frac{\partial M}{\partial X}\right) = -R(M^2 - Q^2 - N_i^2)$$

where $N_i^2 = (\mu_n\mu_p)n_i^2/m_0^2$. Gauss' equation becomes

$$\frac{\partial F}{\partial X} = Q \cosh \beta + M \sinh \beta - Q_c \quad \text{where} \quad \exp \beta = (\mu_n/\mu_p)^{1/2}$$

In classical semiconductor theory R is very small while in this theory R is so large that one can say

$$M^2 - Q^2 \simeq N_i^2 \qquad\qquad \text{A.5.1}$$

This gives the relationship of a hyperbola (Fig. A.4) where the apex A is the point of *maximal resistivity* for the material.

Suppose now that one has a long bar of n-type material with equilibrium concentration n_0 for the electrons and p_0 ($\ll n_0$) for the holes. If into this

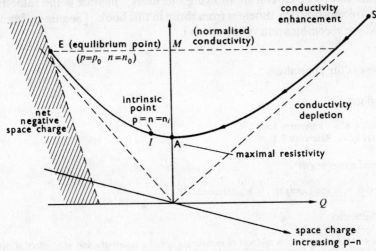

Fig. A.4 Normalized relaxation hyperbola for material where $\mu_n > \mu_p$ (n-type material). Excess holes injected at some point move material to S. Field increases and Q has to decrease, so that M decreases and moves material into a lower conductivity region towards A the maximum possible resistivity, where $\mu_p p = \mu_n n$ but $np = n_i^2$.

bar one injects a large number of holes then, assuming equilibrium is statistically brought about by equilibrium before electrical neutrality, then the concentrations will move directly on to the relaxation hyperbola to some point S. However the charge at this point will be positive, with both M and Q positive and Q_c a negative number. The electric field thus has to rise with distance. One may expect the field close to the injecting contact to be low, so that the current J there will at first have to be carried by the diffusion current $-\partial Q/\partial X$ and Q will fall. Nevertheless, from Gauss, F must continue to rise. Thus the equation for J, $[MF - \partial Q/\partial X]$, is consistent with eqn A.5.1 and Q will move towards the origin while M will decrease towards the maximal resistivity point A.

How the material finally reaches a condition of equilibrium at the point E must be left for further reading.[A.3] The details are not easy, but it can be seen that it cannot just continue with Q decreasing still further and M

rising again, following the relaxation hyperbola. This would be incompatible with the relationship for J, the constant current, unless F was to fall. The field cannot decrease, however, until we have moved past the point where $p - n - p_0 + n_0$ becomes negative. This means that Q has to move around past the equilibrium point E. A complete impass results unless non-equilibrium states are allowed. The essential point emphasized is the need to check that approximations are not carried into situations where they cannot apply. Relaxation effects in semiconductors are good examples of how one has to be careful in applying old ideas; physics is the same but results are substantially different from those in this book. [See also reference A.5, where recombination is important].

References for appendices

Hall effect

SMITH, R. A. Reference 3.2.
KITTEL, C. Reference 2.1.

Material assessment

KANE, P. F. and LARRABEE, G. B. Reference 2.9.

Miscellaneous

A.1 VAN DER PAUW, L. J. A method of measuring specific resistivity and Hall effect of discs of arbitrary shape. *Philips Research Reports*, **13** (1958), 1–9.
A.2 LADBROOKE, P. H. and CARROLL, J. E. The use of a rectangular current pulse for tracing semiconductor resistivity profiles. *Int. J. Electronics*, **31** (1971), 149–172.
A.3 VAN ROOSBROECK, W. and CASEY, H. C. Transport in relaxation semiconductors. *Phys. Rev.*, **5** (1972), 2154–2175.
A.4 ADLER, D. and MOSS, S. C. Amorphous memories and bistable switches. *J. Vacuum Sci. Technol.*, **9** (1972), 1182–1189.
A.5 LAMPERT, M. A. and MARK, P. *Current Injection in Solids*. Academic Press, 1968.

Index